"十四五"职业教育江苏省规划教材

Basic Rubber Technology
橡胶工艺技术概论
（英汉双语）

张小萍　丛后罗　主编
刘琼琼　赵桂英　主审

化学工业出版社
·北京·

本书为英汉双语教材，以橡胶加工工艺基本知识为内容主体，节选相关英文原版经典课文，语言标准、专业性强。全书以英文为主，对课文中的专业词汇、短语及重难点句子进行了中文注释翻译，方便学习者对专业知识内容的学习理解。全书分为6个部分，包括橡胶发展简史（Introduction）、生胶原材料（Introduction to Raw Rubber）、橡胶配合（Introduction to Rubber Compounding）、橡胶加工工艺（Processing Methods in the Rubber Industry）、橡胶硫化（Vulcanization of Rubber）和胶料物理机械性能测试（Property Tests of Uncured and Cured Rubber）。

本书可作为高等职业院校具有橡胶工程特色的相关专业的双语教学教材，也可作为从事橡胶行业的工程技术人员的学习参考书。

图书在版编目（CIP）数据

橡胶工艺技术概论：汉、英/张小萍，丛后罗主编. —北京：化学工业出版社，2019.1（2023.6重印）
ISBN 978-7-122-33339-1

Ⅰ.①橡… Ⅱ.①张…②丛… Ⅲ.①橡胶加工-工艺学-汉、英 Ⅳ.①TQ330.1

中国版本图书馆 CIP 数据核字（2018）第 270357 号

责任编辑：提　岩　于　卉　　　　　　文字编辑：向　东
责任校对：宋　夏　　　　　　　　　　　装帧设计：王晓宇

出版发行：化学工业出版社（北京市东城区青年湖南街13号　邮政编码100011）
印　　装：涿州市般润文化传播有限公司
787mm×1092mm　1/16　印张11　字数294千字　2023年6月北京第1版第2次印刷

购书咨询：010-64518888　　售后服务：010-64518899
网　　址：http://www.cip.com.cn
凡购买本书，如有缺损质量问题，本社销售中心负责调换。

定　　价：36.00元　　　　　　　　　　　　　　　　　　　　　　版权所有　违者必究

前言

随着教学改革的不断深入，各高校相继开设了高分子材料及相关专业的双语教学课程，目的在于培养学生的国际化视野，增强国际交流。在教学过程中，首先面临的问题就是教材的选择。目前，关于高分子材料与工程、材料科学与工程的专业英语书籍很多，但是适合中高职院校及企业人员学习参考的橡胶专业双语教材较少。在此背景下，编者完成了本书的编写。本书中的课文全部选用英文原版经典课文，语言标准、专业性强，在内容上力求反映橡胶工艺技术的知识内容体系及最新发展，使学习者真正领会英文原版课文中专业知识的精髓。同时，本书保留了原课文中的英制单位，读者也可自行换算。

本书分为6个部分，共44课。PART 1 为橡胶发展简史，介绍了天然橡胶、合成橡胶的发展简史；PART 2 为生胶原材料，介绍了生胶的结构、性能及应用；PART 3 为橡胶配合，介绍了配合基本原理及生胶的配合体系（硫化体系、填充补强体系、软化增塑体系及防护体系）；PART 4 为橡胶加工工艺，介绍了混炼、挤出、压延和成型基本知识；PART 5 为橡胶硫化，介绍了硫黄硫化、非硫黄硫化、动态硫化等基本知识；PART 6 为胶料物理机械性能测试，介绍了硫化特性、拉伸性能、耐介质性等胶料常规性重要物性测试基本知识。全书从生胶结构与性能、配合与加工到物理机械性能检测，形成了一套较完整的橡胶加工工艺基本知识体系，便于学生循序渐进地学习。在每篇课文中，对专业性较强及较难理解的单词和短语给出了中文注释，课后给出了专业及重点词汇的音标注释、难点句子的中文注释，帮助学生对专业词汇及课文内容的学习理解。课后的练习题和阅读材料，有助于知识拓展和能力提升。

本书编写分工情况为：PART 1，PART 3，PART 4，PART 6 由徐州工业职业技术学院张小萍编写；PART 2 由徐州工业职业技术学院丛后罗、姚亮编写；PART 5 由丛后罗编写。全书由张小萍、丛后罗担任主编，徐州工业职业技术学院刘琼琼、赵桂英担任主审。在编写过程中还得到了徐州工业职业技术学院材料工程学院的各位老师的帮助，在此深表谢意。

由于编者学术水平、英语水平有限，以及资料的收集、内容的取舍等方面的限制，书中不足之处在所难免，恳请各位读者不吝指正，我们将不断完善和改进。

编者

2018 年 10 月

目录

PART 1　Introduction ⋯⋯⋯⋯⋯ 001
　Lesson 1　History of Natural Rubber ⋯⋯ 001
　Lesson 2　History of Synthetic
　　　　　　Rubber ⋯⋯⋯⋯⋯⋯⋯⋯ 005
**PART 2　Introduction to Raw
　　　　　Rubber** ⋯⋯⋯⋯⋯⋯⋯ 010
　Lesson 3　Natural Rubber (NR) ⋯⋯⋯⋯ 010
　Lesson 4　Polyisoprene Rubber (IR) ⋯⋯ 014
　Lesson 5　Styrene Butadiene Rubber
　　　　　　(SBR) ⋯⋯⋯⋯⋯⋯⋯⋯ 017
　Lesson 6　Polybutadiene Rubber (BR) ⋯⋯ 020
　Lesson 7　Isobutylene-Based Elastomer
　　　　　　(IIR) ⋯⋯⋯⋯⋯⋯⋯⋯⋯ 023
　Lesson 8　Ethylene Propylene Rubber
　　　　　　(EPR) ⋯⋯⋯⋯⋯⋯⋯⋯ 026
　Lesson 9　Nitrile Rubber (NBR) ⋯⋯⋯⋯ 030
　Lesson 10　Neoprene (CR) ⋯⋯⋯⋯⋯⋯ 033
　Lesson 11　Chlorinated and Chlorosulfonated
　　　　　　 Polyethylene (CM, CSM) ⋯⋯ 036
　Lesson 12　Silicone Rubber (MQ,
　　　　　　 MVQ, MPVQ) ⋯⋯⋯⋯⋯⋯ 040
　Lesson 13　Fluorocarbon Rubber
　　　　　　 (FKM) ⋯⋯⋯⋯⋯⋯⋯⋯ 044
　Lesson 14　Thermoplastic Elastomers
　　　　　　 (TPE) ⋯⋯⋯⋯⋯⋯⋯⋯ 047
**PART 3　Introduction to Rubber
　　　　　Compounding** ⋯⋯⋯⋯⋯ 051
　Lesson 15　Introduction to Rubber
　　　　　　 Compounding ⋯⋯⋯⋯⋯ 051
　Lesson 16　Compound Formulation ⋯⋯ 055
　Lesson 17　Raw gum elastomer ⋯⋯⋯ 058
　Lesson 18　Vulcanizing system ⋯⋯⋯⋯ 062
　Lesson 19　Fillers ⋯⋯⋯⋯⋯⋯⋯⋯⋯ 067
　Lesson 20　Plasticizers ⋯⋯⋯⋯⋯⋯⋯ 072
　Lesson 21　Antidegradants ⋯⋯⋯⋯⋯⋯ 075
**PART 4　Processing Methods in the
　　　　　Rubber Industry** ⋯⋯⋯⋯⋯ 079
　Lesson 22　Introduction of Rubber
　　　　　　 Processing ⋯⋯⋯⋯⋯⋯ 079
　Lesson 23　Rubber Mixing ⋯⋯⋯⋯⋯⋯ 083
　Lesson 24　Rubber Extrusion ⋯⋯⋯⋯⋯ 089
　Lesson 25　Calendering of Rubber ⋯⋯⋯ 094
　Lesson 26　Compression Molding ⋯⋯⋯ 098
　Lesson 27　Injection Molding of
　　　　　　 Rubber ⋯⋯⋯⋯⋯⋯⋯⋯ 104
PART 5　Vulcanization of Rubber ⋯⋯⋯ 109
　Lesson 28　Definition ⋯⋯⋯⋯⋯⋯⋯⋯ 109
　Lesson 29　Measurement of
　　　　　　 Vulcanization ⋯⋯⋯⋯⋯⋯ 112
　Lesson 30　Effects of Vulcanization
　　　　　　 on Rubber Properties ⋯⋯⋯ 116
　Lesson 31　Vulcanization with
　　　　　　 Sulfur ⋯⋯⋯⋯⋯⋯⋯⋯⋯ 121
　Lesson 32　Non-Sulfur
　　　　　　 Vulcanization ⋯⋯⋯⋯⋯⋯ 125
　Lesson 33　Dynamic Vulcanization ⋯⋯⋯ 128
**PART 6　Property Tests of Uncured
　　　　　and Cured Rubber** ⋯⋯⋯⋯ 131
　Lesson 34　Introduction of the Basic
　　　　　　 Concepts of Testing ⋯⋯⋯ 131
　Lesson 35　Selected Compound
　　　　　　 Properties and Test ⋯⋯⋯⋯ 135
　Lesson 36　Viscosity Tests ⋯⋯⋯⋯⋯⋯ 138
　Lesson 37　Vulcanization Testing ⋯⋯⋯ 143
　Lesson 38　Tensile Test ⋯⋯⋯⋯⋯⋯⋯ 147
　Lesson 39　Durometer Hardness ⋯⋯⋯⋯ 151
　Lesson 40　Compression Set ⋯⋯⋯⋯⋯ 154
　Lesson 41　Abrasion Resistance ⋯⋯⋯⋯ 158
　Lesson 42　Fluid Resistance ⋯⋯⋯⋯⋯ 161
　Lesson 43　Heat Resistance ⋯⋯⋯⋯⋯ 165
　Lesson 44　Ozone Resistance ⋯⋯⋯⋯ 168
References ⋯⋯⋯⋯⋯⋯⋯⋯⋯⋯⋯⋯ 171

PART 1

Introduction

Lesson 1 History of Natural Rubber

From Columbus onward, European explorers of Central and South America found the natives exploiting the elastic and water resistant properties of the dried latex from certain trees. The indigenous peoples already knew how to crudely waterproof fabrics and boots by coating them with latex and then drying. They also rolled dried latex into the bouncing balls used for sport. This dried latex became quite a curiosity in Europe, especially among the natural scientists. It received the name "rubber" in 1770 when John Priestly discovered that it could rub out pencil marks.

By the early nineteenth century, rubber was recognized as a flexible, tough, waterproof, and air-impermeable material. Commercial exploitation, however, was stymied by the fact that its toughness and elasticity made it difficult to process. More importantly, articles made from it became stiff and hard in cold weather, and soft and sticky in hot weather. The quest to make useful goods from rubber led Thomas Hancock of Great Britain to invent the rubber band, and in 1820, a machine to facilitate rubber processing. His "masticator" subjected the rubber to intensive shearing that softened it sufficiently to allow mixing and shaping. This development was followed, in 1839, by the discovery of vulcanization, which is generally credited to both Hancock and Charles Goodyear of the United States. [1] Vulcanization heating an intimate mixture of rubber and sulfur to crosslink the rubber polymer network greatly improved rubber strength and elasticity and eliminated its deficiencies at temperature extremes. Upon this mechanical and chemical foundation, the rubber industry was born.

The source of rubber latex at that time was the *Hevea brasiliensis* tree (see Figure 1.1), which is native to the Amazon valley. [2] Brazil became the primary source of rubber, but as rubber use grew questions

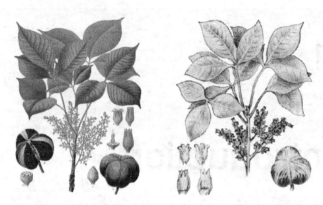

Figure 1.1 Leaf and seed from the *Heveu bruziliensis* tree

arose as to this country's ability to insure adequate supply from its wild rubber trees. In 1876, Henry Wickham collected 70000 Hevea seeds in Brazil and sent them to Kew Gardens in London for germination. Few seedlings resulted, but those that did allowed the British to establish a <u>plantation system</u> throughout the Far East. 种植系统

[3] Dunlop's patenting of the <u>pneumatic tire</u> in England in 1888 ushered in the age of the bicycle as a prelude to the <u>era of automobiles</u>. Tires need rubber, and the demand grew sufficiently great that in the early years of the twentieth century all sources of wild rubber in tropical America and Africa were being tapped. This demand, and the higher prices it caused, turned the plantations in Ceylon (Sri Lanka), Malaya (Malaysia), Singapore, and the East Indies (Indonesia) into prosperous enterprises. By 1914, plantation rubber had overtaken the production of wild rubber, and by 1920 it accounted for 90% of the world's supply.

充气轮胎
汽车时代

New Words

natives	['neɪtɪvz]	n.	土著；当地人
exploit	[ɪk'splɔɪt]	v.	开采；开拓
elastic	[ɪ'læstɪk]	adj.	有弹性的
elasticity	[ˌiːlæ'stɪsəti]	n.	弹性
property	['prɒpəti]	n.	性能；特性
latex	['leɪteks]	n.	胶乳；(尤指橡胶树的)橡浆
indigenous	[ɪn'dɪdʒənəs]	adj.	本地的
waterproof	['wɔːtəpruːf]	v.	使防水
fabric	['fæbrɪk]	n.	织物；布
curiosity	[ˌkjʊəri'ɒsəti]	n.	奇物
flexible	['fleksəbl]	adj.	柔性的
tough	[tʌf]	adj.	坚韧的
toughness	['tʌfnəs]	n.	韧性；刚性

waterproof	['wɔːtəpruːf]	adj.	防水的
stiff	[stɪf]	adj.	坚硬的
sticky	['stɪki]	adj.	黏性的
facilitate	[fə'sɪlɪteɪt]	v.	促进，助长
vulcanization	[ˌvʌlkənaɪ'zeɪʃən]	n.	（橡胶的）硫化（过程），硫化
crosslink	['krɒslɪŋk]	n.	交联
germination	[ˌdʒɜːmɪ'neɪʃn]	n.	萌芽
pneumatic	[njuː'mætɪk]	adj./n.	充气的；有气胎的自行车
usher	['ʌʃə]	v.	引，领
automobile	['ɔːtəməbiːl]	n.	<美>汽车
tropical	['trɒpɪkl]	adj.	热带的
prosperous	['prɒspərəs]	adj.	繁荣的；兴旺的
overtake	[ˌəʊvə'teɪk]	v.	追上，赶上

Notes

[1] Vulcanization heating an intimate mixture of rubber and sulfur to crosslink the rubber polymer network greatly improved rubber strength and elasticity and eliminated its deficiencies at temperature extremes. 硫化加热使橡胶和硫黄的均匀混合物交联形成橡胶聚合物网络，极大地改善了橡胶强度和弹性，并消除了橡胶在极端温度下的缺陷。

[2] Brazil became the primary source of rubber, but as rubber use grew questions arose as to this country's ability to insure adequate supply from its wild rubber trees. 巴西成为橡胶的主要生产国，但随着橡胶用途的增加，这个国家没有能力保证橡胶的充足供应。

[3] Dunlop's patenting of the pneumatic tire in England in 1888 ushered in the age of the bicycle as a prelude to the era of automobiles. 1888年，邓禄普在英格兰申请了充气轮胎专利，引领了自行车时代，成为汽车时代的前奏。

Exercises

1. Translate the following two paragraphs into Chinese.

Natural rubber can be defined as the coagulated, dried rubber prepared from latex extracted from the tree species *Hevea brasiliensis*. More than 200 species of plants produce natural rubbers but only one source, *Hevea brasiliensis*, is commercial today. Reasons for its preeminence are its high yield, longevity, and low resin content.

One natural rubber, guayule（银叶橡胶菊）from the tree species *Parthenium argentatum*, was of commercial importance early in this century. Despite promotion in World War II as a substitute for natural rubber it has been unsuccessful commercially. Yields are not high and it does have high resin content. Its main attraction was that it could be grown in the Southwest region of the USA.

2. Put the following words and phrases into English.

| 弹性 | 性能 | 刚（韧）性 | 柔性的 | 混炼 |
| 成型 | 硫化 | 交联 | 胶乳 | 防水的 |

[Reading Material]

What Exactly Does the Word "Rubber" Mean?

Rubber seems to be a fairly straightforward word. The word derives from a South American Indian word, meaning "weeping wood". The dictionary definition of rubber is, a material that when stretched returns quickly to its "approximate original shape". This definition fits the vulcanized material quite well. ASTM Standard D1566-98 has a detailed definition of rubber implying the vulcanized material. Unfortunately the rubber industry tends to be somewhat casual in the use of the term rubber. When a rubber product is made, the primary raw material is a polymer. This polymer often has some elasticity, but not always. It is then mixed with chemicals to make a rubber "compound" which is subsequently vulcanized. This compound is simply a physical mixture of chemicals and indeed a number of ingredients. The industry often calls both the polymer and the uncured compound, "rubber". Unvulcanized silicone, for example, (both polymer and uncured compound) does not fit the dictionary definition too well, since it can have the consistency of butter. The word "rubber" used in the book for the primary raw polymer will be raw gum elastomer.

Some people use the term rubber, to mean NR only, but there have been instances when a customer asked for rubber (expecting the vendor to choose the right type, neoprene, natural, etc.), the vendor however assumed the customer was specifically asking for natural rubber, which in that case turned out to be the worst choice. Naturally, the vulcanized material is also called rubber, as indeed it should be. The word "elastomer" and "rubber" are often used by the industry to mean exactly the same thing, which is a waste of such an interesting word (elastomer), which maybe, could have been reserved exclusively to describe the raw polymer. ASTM Standard D1566-98 defines "elastomer" as "a term often used for rubber and polymers that have properties similar to those of rubber".

Words and Expressions

term	[tɜːm]	n.	术语
stretch	[stretʃ]	v.	伸展；延伸
definition	[ˌdefɪˈnɪʃn]	n.	定义；规定
raw	[rɔː]	adj.	生的，未加工的
compound	[ˈkɒmpaʊnd]	n.	胶料
silicone	[ˈsɪlɪkəʊn]	n.	硅树脂
vendor	[ˈvendə(r)]	n.	供应商

Lesson 2　History of Synthetic Rubber

[1] The ready availability of high quality plantation rubber facilitated the advances in production methods and product quality which catalyzed the development of better automobiles and their expanding reliance on rubber products. The demand for rubber also prompted research into the synthesis of practical substitutes. As early as the 1880s, organic chemists had identified isoprene as the main structural unit of rubber. By 1890, several researchers had made synthetic rubber-like polyisoprene. With wild and plantation rubber readily available, however, synthetic alternatives remained mostly of academic interest. The allied blockade of Germany during World War I changed this by demonstrating to Germany, and the world, the strategic importance of rubber in war. The Germans produced more than 2000 tons of methyl rubber by polymerizing 2,3-dimethyl-1,3-butadiene, but its properties were poor and its production was abandoned at the war's end.

The lesson was learned, nevertheless. [2] The governments of Germany and Russia, the military nations most susceptible to loss of natural rubber through naval blockade, instituted programs to develop synthetic alternatives. This was given additional impetus in the mid-1920s by a forced rise in rubber prices due to British restrictions on plantation production. Work in Germany and Russia concentrated on polymers of 1,3-butadiene, since it was less expensive and easier to manufacture than isoprene (2-methyl-1,3-butadiene). This led to the production in the early 1930s of sodium catalyzed polybutadiene as SK rubber in Russia and Buna (from butadiene and Na) in Germany. These polymers were hard, tough, and difficult to process. [3] Copolymerization of 1,3-butadiene with other monomers was pursued to obtain more easily processed and rubber-like products. In Germany this led within a few years to Buna S, a copolymer of butadiene and styrene, and Buna N, a copolymer of butadiene and acrylonitrile.

[4] By the time war broke out in 1939, both Germany and Russia could satisfy their rubber needs with reasonably satisfactory synthetic products. As the war spread to the Far East, the U.S. government realized its supply of natural rubber was at risk. In 1940 it established the Rubber Reserve Company as a government corporation. This organization was charged with stockpiling natural rubber and instituting a synthetic rubber research and development program. Based on the technology developed in Germany and Russia prior to the war, styrene-butadiene polymers soon became the focus of development efforts as the best general-purpose alternative to natural rubber. Production of styrene-butadiene rubber

(SBR), then called GR S, began in a government plant in 1942. Over the next three years, government financed construction of 15 SBR plants brought annual production to more than 700000 tons.

Although SBR was the workhorse substitute for natural rubber during the war years, it was supplemented by neoprene, butyl rubber, and nitrile rubber. Neoprene, poly(2-chloro-1, 3-butadiene) had been introduced by DuPont in 1931 as DuPrene and was actually the first commercially successful synthetic elastomer. Neoprene provided high tensile strength like natural rubber and significantly better weather and oil resistance, but it was too expensive for consideration as a general purpose elastomer. Its production rose from 9000 tons to 45000 tons between 1942 and 1945. Butyl rubber had been invented at Standard Oil when it was discovered that the saturated chemical structure of rubber-like high molecular weight isobutylene polymers could be vulcanized by incorporating a small amount of isoprene into the polymer. The resulting elastomer showed unique and useful properties, particularly low gas permeability. Construction of the first butyl rubber plant had already begun in 1941, but the facility was nationalized and put into full production by 1943. From a base of 1400 tons in 1943, butyl rubber production rose rapidly to over 47000 tons in 1945. Nitrile rubber, the current common name for acrylonitrile-butadiene rubber, showed such great potential as a special purpose oil-and solvent-resistant rubber that its commercialization in Germany was followed by production at four new U. S. plants by the time of America's entry into the war. By the end of the war, production had risen to 15000 tons per year.

World War Ⅱ represents the accelerated adolescence of the rubber industry, the preparation for its great postwar maturation and growth. Developments in organic and polymer chemistry provided insight into the properties of natural rubber, which led to neoprene, SBR, nitrile and butyl rubber. These were made practically and commercially significant, however, only by the complementary developments in compounding. Amine vulcanization accelerators were discovered at Diamond Rubber in 1906. The value of carbon black to the improvement of rubber physical properties was discovered at Diamond Rubber in 1912. Internal mixers for compounding were first offered by Fernley Banbury in 1916. Guanidines and thiazoles began use as accelerators in 1921. [5] By the end of the Second World War, compounding art and science had fully bridged the gap between elastomer and rubber product.

氯丁橡胶/丁基橡胶
丁腈橡胶/聚 2-氯-1,3-丁二烯
拉伸强度
耐油性
通用橡胶

异丁烯聚合物

低透气性

全面投产

丁腈橡胶
耐油、耐溶剂橡胶

青春期

配合
胺类硫化促进剂
炭黑/物理机械性能
密炼机
胍类
噻唑类

New Words

synthesis	['sɪnθəsɪs]	n.	合成
synthetic	[sɪn'θetɪk]	adj.	合成的

substitute	['sʌbstɪtjuːt]	n.	代替者,替代物
isoprene	['aɪsəpriːn]	n.	异戊二烯
strategic	[strə'tiːdʒɪk]	adj.	战略(上)的;战略性的
polymerize	['pɒlɪməraɪz]	v.	(使)聚合
susceptible	[sə'septəbl]	adj.	易受影响的
butadiene	[ˌbjuːtə'daɪiːn]	n.	丁二烯
polybutadiene	[ˌpɒlɪbjuːtə'daɪiːn]	n.	聚丁二烯
copolymerization	[kəʊpɒlɪməraɪ'zeɪʃən]	n.	共聚合(作用)
copolymer	[kəʊ'pɒlɪmə]	n.	共聚物
monomer	['mɒnəmə]	n.	单体
styrene	['staɪriːn]	n.	苯乙烯
acrylonitrile	[ækrələʊ'naɪtrɪl]	n.	丙烯腈
satisfactory	[ˌsætɪs'fæktəri]	adj.	令人满意的
neoprene	['niːəpriːn]	n.	氯丁橡胶
butyl	['bjuːtɪl]	n.	丁基,丁基合成橡胶
nitrile	['naɪtrɪl]	n.	腈
isobutylene	[aɪsəʊ'bjuːtɪliːn]	n.	异丁烯
permeability	[ˌpɜːmɪə'bɪləti]	n.	渗透性;可渗透性
commercialization	[kəˌmɜːʃəlaɪ'zeɪʃn]	n.	商业化,商品化
maturation	[ˌmætʃu'reɪʃn]	n.	成熟
compounding	['kɒmpaʊndɪŋ]	n.	混合

Notes

[1] The ready availability of high quality plantation rubber facilitated the advances in production methods and product quality which catalyzed the development of better automobiles and their expanding reliance on rubber products. 高品质种植橡胶的随时可用,促进了生产方法和产品质量的进步,催化了更好的汽车工业的发展并扩大了其对橡胶产品的依赖。

[2] The governments of Germany and Russia, the military nations most susceptible to loss of natural rubber through naval blockade, instituted programs to develop synthetic alternatives. This was given additional impetus in the mid-1920s by a forced rise in rubber prices due to British restrictions on plantation production. 德国和俄罗斯这两个最容易因海军封锁而丧失天然橡胶的军事国家制定了开发(天然橡胶)合成替代品的计划。在20世纪20年代中期,由于英国对种植园生产的限制,橡胶价格受到强制上涨,给予(合成橡胶的生产)额外的推动力。

[3] Copolymerization of 1,3-butadiene with other monomers was pursued to obtain more easily processed and rubber-like products. In Germany this led within a few years to Buna S, a copolymer of butadiene and styrene, and Buna N, a copolymer of butadiene and acrylonitrile. 将1,3-丁二烯与其他单体共聚,以获得更容易加工的、更类似于橡胶的产品。在德国,丁二烯与苯乙烯的共聚物丁腈橡胶S、丁二烯与丙烯腈的共聚物丁腈橡胶N在几年之内得到了开发。

[4] By the time war broke out in 1939, both Germany and Russia could satisfy their rubber needs with reasonably satisfactory synthetic products. 到1939年战争爆发时,德国和俄罗斯都可以用相当令人满意的合成橡胶产品来满足他们的橡胶需求。

[5] By the end of the Second World War, compounding art and science had fully bridged the gap between elastomer and rubber product. 到第二次世界大战结束时，配合艺术和科学已经完全弥合了弹性体和橡胶产品之间的差距。

Exercises

1. Translate the following paragraph into Chinese.

Synthetic elastomers were first produced in Germany in the early 1930s with a world consumption of 2500 long tons. After World War II in 1946, the world consumption of synthetic rubber had grown to 912400 long tons in comparison to 555000 long tons of natural rubber.

2. Put the following words and phrases into English.

合成的	合成橡胶	结构单元	单体	聚合
共聚物	拉伸强度	炭黑	配合	耐油橡胶

[Reading Material]

Rubber Appreciation

Rubber is a fascinating, marvelous material with a unique combination of properties. The compounder originates recipe to optimize one, several, or all of the rubber's inherent capabilities to provide a compound that will be molded or formed into the desired useful marketable produce.

A rubber band is an example that utilizes the stretch nature of rubber. A tire tread needs to be primarily flexible and abrasive resistant. A special inner-liner of the tire has amazing resistance to air permeation while rubber boots need to be waterproof. These are only a few examples of the usefulness of rubber; the list is almost endless.

The compounder has to know what materials are available for his purpose. And regardless of how many years of experience he/she can rely on, there is never any end to the accumulation of new knowledge. Moreover, attention has already given to the recipe ingredients to achieve ease and safety of processing behavior of the compound during mixing, extrusion and calendaring in the factory prior to the forming, molding and curing of the product. Premature scorch, set-up or curing during processing, must be avoided.

The historical account of compounding will provide a basis for what is to follow in the text and hopefully cultivate a sense of appreciation among the readers for the art and science of putting together recipe for product performance.

Words and Expressions

fascinating	[ˈfæsɪneɪtɪŋ]	adj.	迷人的；使人神魂颠倒的
recipe	[ˈresəpi]	n.	配方
rubber band			橡皮筋
tire	[ˈtaɪə(r)]	n.	轮胎
tire tread			轮胎胎面
abrasive resistant			耐磨性

inner-liner			内衬
waterproof			防水的
mixing	['mɪksɪŋ]	n.	混炼
extrusion	[ɪk'struːʒn]	n.	挤出
calendaring	['kælɪndərɪŋ]	n.	压延
scorch	[skɔːtʃ]	n.	焦烧
rubber compounding			橡胶配合

PART 2

Introduction to Raw Rubber

Lesson 3　Natural Rubber (NR)

Natural rubber is extracted in the form of latex from the bark of the Hevea tree. The rubber is collected from the latex in a series of steps involving preservation, concentration, coagulation, dewatering, drying, cleaning, and blending. Because of its natural derivation, it is sold in a variety of grades based on purity (color and presence of extraneous matter), viscosity, viscosity stability, oxidation resistance, and rate of cure.

$$\left[CH_2 - \underset{\underset{CH_3}{|}}{C} = CH - CH_2 \right]_n$$

NR/IR

The natural rubber polymer is nearly 100% cis-1,4-polyisoprene with M_w ranging from 1×10^6 to 2.5×10^6. Due to its high structural regularity, natural rubber tends to crystallize spontaneously at low temperatures or when it is stretched. Low temperature crystallization causes stiffening, but is easily reversed by warming. [1] The strain-induced crystallization gives natural rubber products high tensile strength and resistance to cutting, tearing, and abrasion.

Elasticity is one of the fundamentally important properties of natural rubber. [2] Rubber is unique in the extent to which it can be distorted, and the rapidity and degree to which it recovers to its original shape and dimensions. It is, however, not perfectly elastic. The rapid recovery is not complete. Part of the distortion is recovered more slowly and part is retained. The extent of this permanent distortion, called permanent set, depends upon the rate and duration of the applied force. The slower the force, and the longer it is maintained, the greater is the permanent set. Because of rubber's elasticity, however, the permanent set may not be complete even after long periods of applied force. This quality is of obvious value in gaskets and seals.

As already noted, this rubber polymer network was originally an <u>impediment</u> to rubber processing. Mixing additives with a tough, elastic piece of raw rubber was a substantial challenge. The solution came with the discovery of its <u>thermoplastic</u> behavior. High shear and heat turn the rubber soft and plastic. In this state it is considerably more receptive to the incorporation of additives so that the rubber's natural attributes can be modified and optimized as desired. [3] The commercial utility of natural rubber has in fact grown from the ease with which its useful properties can be changed or improved by compounding techniques.

障碍物
热塑性

Another important and almost unique quality of uncured natural rubber compounds is <u>building tack</u>. When two fresh surfaces of milled rubber are pressed together they bond into a single piece. This facilitates the building of composite articles from separate components. In tire manufacture, for example, the separate pieces of uncured tire are held together solely by building tack. During cure they fuse into a single unit.

黏性

<u>Modified natural rubbers</u> are also sold, with treatment usually performed at the latex stage. These include <u>epoxidized natural rubber</u> (ENR); <u>deproteinized natural rubber</u> (DNR); <u>oil extended natural rubber</u> (OENR), into which 10%～40% of process oils have been incorporated; Hevea plus MG rubber—natural rubber with grafted poly (methyl methacrylate) side chains; and <u>thermoplastic natural rubber</u> (TNR)—blends of natural rubber and polypropylene.

改性天然橡胶
环氧化天然橡胶
脱蛋白天然橡胶/充油天然橡胶
热塑性天然橡胶

New Words

preservation	[ˌprezə'veɪʃn]	n.	保存,保留
concentration	[ˌkɒnsn'treɪʃn]	n.	浓度,浓缩
coagulation	[kəʊˌægjʊ'leɪʃn]	n.	凝结
dewatering	[diː'wɔːtərɪŋ]	n.	脱水
derivation	[derɪ'veɪʃn]	n.	衍生物
regularity	[regju'lærəti]	n.	规整性
crystallize	['krɪstəlaɪz]	v.	(使)结晶
spontaneously	[spɒn'teɪnɪəsli]	adv.	自发地
reverse	[rɪ'vɜːs]	v.	(使)反转
abrasion	[ə'breɪʒn]	n.	磨耗
permanent	['pɜːmənənt]	adj.	永久(性)的
distortion	[dɪ'stɔːʃn]	n.	变形
impediment	[ɪm'pedɪmənt]	n.	障碍物;妨碍
epoxidize	[e'pɒksɪdaɪz]	v.	使环氧化
deproteinize	[diːˌprəʊtɪnaɪz]	v.	使脱(去)蛋白质

Notes

[1] The strain-induced crystallization gives natural rubber products high tensile strength and

resistance to cutting, tearing, and abrasion. 拉伸诱导结晶可以提高天然橡胶的拉伸强度、抗剪强度、抗撕裂性能和耐磨性。

[2] Rubber is unique in the extent to which it can be distorted, and the rapidity and degree to which it recovers to its original shape and dimensions. 橡胶可以发生形变,并且能够迅速恢复形状和尺寸的特性是独一无二的。

[3] The commercial utility of natural rubber has in fact grown from the ease with which its useful properties can be changed or improved by compounding techniques. 实际上随着配合技术的提高,天然橡胶的性能得到了改善,进而扩大天然橡胶的商业应用。

Exercises

1. Translate the first and second paragraphs of the text into Chinese.
2. Put the following words into English.

| 天然橡胶 | 环氧化天然橡胶 | 结晶 | 耐磨性 | 脱水 |
| 胶乳 | 弹性 | 永久变形 | 拉伸诱导结晶 | |

[Reading Material]

Properties of NR

Vulcanized products made from NR have high mechanical strength and can be compounded to have excellent elasticity (ability to snap back to their original shape). NR has very good abrasion resistance which, with its low relative cost, makes it a significant choice for slurry pump liners and impellers as well as for tank linings. It has very good dynamic mechanical properties and is therefore used in tires, rubber springs and vibration mounts. It is one of the few elastomers that have high strength. As NR rubber gum vulcanizate has a very high elasticity, thus most of the kinetic energy of an impacting particle is converted into deformation of the vulcanizate which then releases the energy by returning to its original undeformed state.

It also has very good low temperature resistance, down into the region of $-57℃$ at which its stiffness shows a considerable increase. Its high temperature heat aging resistance limit for continuous use is in the region of 75℃. Inherent weather resistance provided by the raw elastomer is poor. Significant components of weather, from the rubber technologists point of view, are UV light and ozone. Addition of carbon black to a compound gives resistance to UV, antiozonants and waxes, and helps with ozone resistance. Ozone attack is of most concern for thin products and those that are subjected to stretching in service.

Words and Expressions

slurry pump			泥浆泵
impeller	[ɪmˈpelə]	n.	叶轮
dynamic mechanical property			动态力学性能特性
vibration mount			橡胶减振器

stiffness	[ˈstɪfnəs]	n.	硬度
UV light			紫外光
ozone	[ˈəʊzəʊn]	n.	臭氧
antiozonant	[æntiˈəʊzəʊnənt]	n.	抗臭氧剂

Lesson 4 Polyisoprene Rubber (IR)

Polyisoprene is made by solution polymerization of isoprene (2-methyl-1,3-butadiene). [1] The isoprene monomer, the structural unit of the natural rubber polymer, can polymerize in four isomeric forms: *trans*-1,4- addition, *cis*-1,4-addition, 1,2-addition, leaving a pendant vinyl group, and 3,4-addition. The production of a synthetic analogue to natural rubber was stymied for over 100 years because polymerization of isoprene resulted in mixtures of isomeric forms. In the 1950s, rubber-like elastomers with $>90\%$ *cis*-1,4-isoprene configuration were finally produced using stereospecific catalyst.

Polyisoprene compounds, like those of natural rubber, exhibit good building tack, high tensile strength, good hysteresis, and good hot tensile and hot tear strength. The characteristics which differentiate polyisoprene from natural rubber arise from the former's closely controlled synthesis. Polyisoprene is chemically purer—it does not contain the proteins and fatty acids of its natural counterpart. Molecular weight is lower than natural rubber's, and lot-to-lot uniformity is better. [2] Polyisoprene is therefore easier to process, gives a less variable (although generally slower) cure, is more compatible in blends with EPDM and solution SBR, and provides less green strength (pre-cure) than natural rubber. Polyisoprene is added to SBR compounds to improve tear strength, tensile strength, and resilience while decreasing heat buildup. Blends of polyisoprene and fast curing EPDM combine high ozone resistance with the good tack and cured adhesion uncharacteristic of EPDM alone.

Polyisoprene is typically used in favor of natural rubber in applications requiring consistent cure rates, tight process control, or improved extrusion, molding, and calendering. Tires are the leading consumer. [3] The synthetic elastomer can be produced with the very low level of branching, high molecular weight, and relatively narrower molecular weight distribution that contributes to lower heat buildup compared to natural rubber. For this reason, certain grades of polyisoprene are used as an alternative to natural rubber in the tread of high service tires (truck, aircraft, off-road) without sacrificing abrasion resistance, groove cracking, rib tearing, cold flex properties, or weathering resistance. Footwear and mechanical goods are also major uses. Because of polyisoprene's high purity and the high gum (unfilled) tensile strength of its compounds, it is widely used in medical goods and food-contact items. These include baby bottle nipples, milk tubing, and hospital sheeting.

聚异戊二烯/溶液聚合单体	
同分异构的	
合成的相似物	
构象	
有规立构催化剂	
滞后现象	
化学纯	
批次	
格林强度	
生热	
黏附力	
硫化速率	
分子量分布	
骨架开裂/低温屈挠	
食品级的	

New Words

polyisoprene	[pɒlɪˈaɪsəupriːn]	n.	聚异戊二烯
solution	[səˈluːʃn]	n.	溶液
monomer	[ˈmɒnəmə]	n.	单体
isomeric	[ˌaɪsəˈmerɪk]	adj.	异构的
synthetic	[sɪnˈθetɪk]	adj.	合成的
analogue	[ˈænəlɒg]	n.	相似物
configuration	[kənˌfɪgəˈreɪʃn]	n.	构象
catalyst	[ˈkætəlɪst]	n.	催化剂
hysteresis	[ˌhɪstəˈriːsɪs]	n.	磁滞现象
differentiate	[ˌdɪfəˈrenʃɪeɪˌet]	v.	区别
counterpart	[ˈkaʊntərpɑːrt]	n.	副产物,杂质
uniformity	[ˌjuːnɪˈfɔːmətɪ]	n.	均一性
compatible	[kəmˈpætəbəl]	adj.	相容的
adhesion	[ædˈhiːʒən]	n.	黏合力,黏附力
uncharacteristic	[ˌʌnkærəktəˈrɪstɪk]	adj.	不典型的
groove	[gruːv]	n.	沟,槽
purity	[ˈpjʊrɪtɪ]	n.	纯度

Notes

[1] The isoprene monomer, the structural unit of the natural rubber polymer, can polymerize in four isomeric forms: *trans*-1,4-addition, *cis*-1,4-addition, 1,2-addition, leaving a pendant vinyl group, and 3,4-addition. 异戊二烯单体聚合得到的结构单元,即天然橡胶的结构单元,有四种不同的异构体:反-1,4-结构、顺-1,4-结构、带有乙烯基侧链的1,2结-构和3,4-结构。

[2] Polyisoprene is therefore easier to process, gives a less variable (although generally slower) cure, is more compatible in blends with EPDM and solution SBR, and provides less green strength (pre-cure) than natural rubber. 因此,聚异戊二烯橡胶更容易加工,表现出重现性很好的硫化特性(尽管比天然橡胶稍慢),还能与EPDM和溶聚SBR良好地相容,也比天然橡胶的格林强度低。

[3] The synthetic elastomer can be produced with the very low level of branching, high molecular weight, and relatively narrower molecular weight distribution that contributes to lower heat buildup compared to natural rubber. 这种合成的弹性体具有较小的支化程度,较高的分子量和更窄的分子量分布,这些结构特点使得生热会比天然橡胶低。

Exercises

1. Translate the first and second paragraphs of the text into Chinese.
2. Put the following words into English.

聚异戊二烯橡胶	有规立构的	溶液	合成的	催化剂
纯度	分子量分布			

[Reading Material]

Application of IR

Polyisoprene elastomers (IR) are synthetic, predominantly stereoregular polymers that closely resemble natural rubber in molecular structure as well as in properties. Polyisoprene elastomers are currently being used in a variety of applications requiring good resilience, low water swell, high gum tensile strength, good tack and high hot tensile strength. The largest end use for polyisoprene by far is in tires. Black-loaded polyisoprene finds uses in tires, motor mounts, pipe gaskets, shock absorber bushings, and many other molded and mechanical goods. Gum polyisoprene compounds are used in rubber bands, cut thread, baby bottle nipples, and extruded hoses, and other such items. Mineral-filled polyisoprene finds applications in footwear, sponges, and sporting goods. Other important uses include medical applications and adhesives and sealants. Of all the applications, medical applications have seen the fastest growth.

Words and Expressions

stereoregular	[stɪərɪəˈreɡjʊlə]	adj.	有规立构的
resilience	[rɪˈzɪlɪəns]	n.	弹性
shock absorber			减振器
sponge	[spʌndʒ]	n.	海绵
sealant	[ˈsiːlənt]	n.	密封剂

Lesson 5 Styrene Butadiene Rubber (SBR)

SBR polymers are widely produced by both emulsion and solution polymerization. [1] Emulsion polymerization is carried out either hot, at about 50℃, or cold, at about 5℃, depending upon the initiating system used. SBR made in emulsion usually contains about 23% styrene randomly dispersed with butadiene in the polymer chains. SBR made in solution contains about the same amount of styrene, but both random and block copolymers can be made. Block styrene is thermoplastic and at processing temperatures helps to soften and smooth out the elastomer. Both cold emulsion SBR and solution SBR are offered in oil-extended versions. These have up to 50% petroleum base oil on polymer weight incorporated within the polymer network. Oil extension of SBR improves processing characteristics, primarily allowing easier mixing, without sacrificing physical properties.

$$\text{+}(CH_2-CH=CH-CH_2)_5\text{-}(\overset{C_6H_5}{\underset{|}{CH}}-CH_2)\text{+}_n$$
SBR

[2] SBR was originally produced by the hot emulsion method, and was characterized as more difficult to mill, mix, or calender than natural rubber, deficient in building tack, and having relatively poor inherent physical properties. Processability and physical properties were found to be greatly improved by the addition of process oil and reinforcing pigments. "Cold" SBR generally has a higher average molecular weight and narrower molecular weight distribution. It thereby offers better abrasion and wear resistance plus greater tensile and modulus than "hot" SBR. [3] Since higher molecular weight can make cold SBR more difficult to process, it is commonly offered in oil-extended form. Solution SBR can be tailored in polymer structure and properties to a much greater degree than their emulsion counterparts. The random copolymers offer narrower molecular weight distribution, low chain branching, and lighter color than emulsion SBR. They are comparable in tensile, modulus, and elongation, but offer lower heat buildup, better flex, and higher resilience. Certain grades of solution SBR even address the polymer's characteristic lack of building tack, although it is still inferior to that of natural rubber.

The processing of SBR compounds in general is similar to that of natural rubber in the procedures and additives used. SBR is typically compounded with better abrasion, crack initiation, and heat resistance than natural rubber. SBR <u>extrusions</u> are smoother and maintain their shape better than those of natural rubber. 挤出制品

SBR was originally developed as a <u>general purpose elastomer</u> and it still retains this <u>distinction</u>. It is the largest volume and most widely used elastomer worldwide. Its single largest application is in <u>passenger car tires</u>, particularly in tread compounds for <u>superior traction</u> and <u>tread wear</u>. Substantial quantities are also used in footwear, <u>foamed products</u>, wire and cable jacketing, belting, hoses, and mechanical goods. 通用弹性体 / 区别 / 乘用车轮胎 / 卓越牵引力/胎面磨损 / 泡沫制品

New Words

emulsion	[ɪˈmʌlʃən]	n.	乳液
initiate	[ɪˈnɪʃɪeɪt]	v.	引发
random	[ˈrændəm]	adj.	无规的,随机的
petroleum	[pəˈtrəʊlɪəm]	n.	石油
primarily	[praɪˈmerəli]	adv.	根本上
calender	[ˈkælɪndə]	v.	压延
processability	[prəʊsesəˈbɪlɪtɪ]	n.	加工性能
modulus	[ˈmɒdjʊləs]	n.	模量
procedure	[prəˈsiːdʒə(r)]	n.	程序,过程
extrusion	[ɪkˈstruːʒn]	n.	挤出
distinction	[dɪˈstɪŋkʃn]	n.	区别
traction	[ˈtrækʃn]	n.	附着力

Notes

[1] Emulsion polymerization is carried out either hot, at about 50℃, or cold, at about 5℃, depending upon the initiating system used. 乳液聚合可以在高温（约50℃）下进行，也可以在低温（约5℃）下进行，这取决于所使用的引发体系。

[2] SBR was originally produced by the hot emulsion method, and was characterized as more difficult to mill, mix, or calender than natural rubber, deficient in building tack, and having relatively poor inherent physical properties. 最初SBR是用高温乳液聚合的方法进行生产的，但是它比天然橡胶更难混炼和压延，而且加工黏性差、力学强度低。

[3] Since higher molecular weight can make cold SBR more difficult to process, it is commonly offered in oil-extended form. 因为较高的分子量使得低温SBR难以加工，所以通常低温SBR需要充油以改善其加工性能。

Exercises

1. Translate the first and second paragraphs of the text into Chinese.
2. Put the following words and phrases into English.

丁苯橡胶　　　充油丁苯橡胶　　　乳液聚合　　　模量　　　泡沫制品
无规共聚物　　补强　　　　　　乳聚丁苯橡胶　　溶聚丁苯橡胶

[Reading Material]

Emulsion-SBR and Solution-SBR

SBR is derived from two monomers, styrene and butadiene. The mixture of these two monomers is polymerized by two processes: from solution or as an emulsion. Emulsion SBR (E SBR) produced by emulsion polymerization is initiated by free radicals. Reaction vessels are typically charged with the two monomers, a free radical generator, and a chain transfer agent such as an alkyl mercaptan. Radical initiators include potassium persulfate and hydroperoxides in combination with ferrous salts. Emulsifying agents include various soaps. By capping the growing organic radicals, mercaptans, control the molecular weight, and hence the viscosity, of the product. Typically, polymerizations are allowed to proceed only to ca. 70%, a method called "short stopping". In this way, various additives can be removed from the polymer.

Solution SBR (S SBR) is produced by an anionic polymerization process. Polymerization is initiated by alkyl lithium compounds. Water is strictly excluded. The process is homogeneous, which provides greater control over the process, allowing tailoring of the polymer. The alkyl lithium compound adds to one of the monomers, generating a carbanion that then adds to another monomer, and so on. Relative to E SBR, S SBR is increasingly favored because it offers improved wet grip and rolling resistance, which translate to greater safety and better fuel economy, respectively.

Words and Expressions

reaction vessel			反应釜
free radical			自由基
generator	['dʒɛnə,retə]	n.	引发剂
chain transfer agent			链转移剂
alkyl	['ælkɪl]	n.	烷基
mercaptan	[mə'kæptæn]	n.	硫醇
potassium persulfate			过硫酸钾
hydroperoxide	[haɪdroʊpə'rɒksaɪdz]	n.	氢过氧化物
ferrous salt		n.	亚铁盐
emulsifying agent			乳化剂
viscosity	[vɪ'skɑsɪti]	n.	黏度
anionic polymerization			阴离子聚合物
alkyl lithium			烷基锂
homogeneous	[,hoʊmə'dʒiːniəs]	n.	均相的
carbanion	[kɑːbæn,aɪɒn]	n.	碳负离子
wet grip resistance			抗湿滑性

Lesson 6　Polybutadiene Rubber (BR)

Polybutadiene elastomer was originally made by emulsion polymerization, generally with poor results. It was difficult to process and did not extrude well. Polybutadiene became commercially successful only after it was made by solution polymerization using stereospecific Ziegler-Natta catalysts. This provided a polymer with greater than 90% *cis*-1,4-polybutadiene configuration. This structure hardens at much lower temperatures (with T_g of $-100℃$) than natural rubber and most other commercial elastomers. This gives better low temperature flexibility and higher resilience at ambient temperatures than most elastomers. Greater resilience means less heat buildup under continuous dynamic deformation as well. [1] This high-cis BR was also found to possess superior abrasion resistance and a great tolerance for high levels of extender oil and carbon black. [2] High-cis BR was originally blended with natural rubber simply to improve the latter's processing properties, but it was found that the BR conferred many of its desirable properties to the blend. The same was found to be true in blends with SBR.

$$\pm CH_2-CH=CH-CH_2\pm_n$$
BR

The 1,3-butadiene monomer can polymerize in three isomeric forms: by *cis*-1,4-addition, *trans*-1,4-addition, and 1,2-addition leaving a pendant vinyl group. [3] By selection of catalyst and control of processing conditions, polybutadienes are now sold with various distributions of each isomer within the polymer chain, and with varying levels of chain linearity, molecular weight and molecular weight distribution. Each combination of chemical properties is designed to enhance one or more of BR's primary attributes.

The largest volume use of polybutadiene is in passenger car tires, primarily in blends with SBR or natural rubber to improve hysteresis (resistance to heat buildup), abrasion resistance, and cut growth resistance of tire treads. The type of BR used depends on which properties are most important to the particular compound. High-cis and medium-cis BRs have excellent abrasion resistance, low rolling resistance, but poor wet traction. High-vinyl BRs offer good wet traction and low rolling resistance, but poor abrasion resistance. Medium-vinyl BR balance reasonable wet traction with good abrasion resistance and low rolling resistance. Polybutadiene is also used for improved durability and abrasion and flex crack resistance in tire sidewalls, as well as in elastomer blends for belting. High-cis and medium-cis BRs are also used in the manufacture of

齐格勒-纳塔催化剂

环境温度
动态的
耐磨性

工艺性能

工艺条件

异构体
分子量及分子量分布

轮胎胎面

滚动阻力

耐久性
胎侧

high impact polystyrene. 3%～12% BR is grafted onto the styrene chain as it polymerizes, conferring high impact strength to the resultant polymer. | 高抗冲聚苯乙烯

New Words

extrude	[ɪk'struːd]	v.	挤出
ambient	['æmbiənt]	adj.	环境；周围的
dynamic	[daɪ'næmɪk]	adj.	动态的
deformation	[diːfɔːˈmeɪʃn]	n.	变形
confer	[kən'fɜː(r)]	v.	授予，颁予
distribution	[ˌdɪstrɪ'bjuːʃn]	n.	分配，分布
linearity	[ˌlɪnɪ'ærətɪ]	n.	线型
primary	[praɪməri]	adj.	主要的；基本的
attribute	[ə'trɪbjuːt]	v.	把…归于
durability	[ˌdjʊərə'bɪlətɪ]	n.	耐久性
sidewall	['saɪdwɔːl]	n.	侧墙

Notes

[1] This high-cis BR was also found to possess superior abrasion resistance and a great tolerance for high levels of extender oil and carbon black. 高顺式聚丁二烯橡胶具有优异的耐磨性，另外它还具有油和炭黑的高填充性。

[2] High-cis BR was originally blended with natural rubber simply to improve the latter's processing properties, but it was found that the BR conferred many of its desirable properties to the blend. 高顺式聚丁二烯橡胶最初和天然橡胶共混，可以明显提高天然橡胶的工艺性能，后来发现共混顺丁橡胶能够改善胶料很多性能。

[3] By selection of catalyst and control of processing conditions, polybutadienes are now sold with various distributions of each isomer within the polymer chain, and with varying levels of chain linearity, molecular weight and molecular weight distribution. 通过选择不同的催化剂和控制工艺条件，可以得到不同结构单元、线型结构、分子量及其分布的聚丁二烯大分子。

Exercises

1. Translate the first and second paragraphs of the text into Chinese.
2. Put the following words into English.

| 聚丁二烯橡胶 | 顺丁橡胶 | 耐久性 | 动态形变 | 滚动阻力 |
| 高抗冲聚苯乙烯 | 工艺性能 | 顺-1,4-加成 | | |

[Reading Material]

BR Blends

Although BR is a significant elastomer it is most commonly used as a blend with other rubbers. Grades are very much dependent on the architecture of the repeating unit in the polymer chain. BR is traditionally difficult to process on rubber machinery; this difficulty is not apparent when BR is blended with other non-polar elastomers such as NR. BR vulcanizates confer high resilience, therefore low heat buildup, and good abrasion resistance to blends with other rubbers (its resilience is excellent and it has a low temperature flexibility second only to silicone rubber). In view of the above properties its major application area is in tires. Other applications are golf ball centers, modification of polystyrene to make high impact polystyrene and miscellaneous products needing improvements in abrasion, low temperature and resilience.

Words and Expressions

grade	[greɪd]	n.	等级
architecture	[ˈɑːkɪtektʃə(r)]	n.	结构
repeating unit			重复单元
non polar elastomers			非极性弹性体
vulcanizate	[ˈvʌlkənɪzeɪt]	n.	硫化产品
flexibility	[ˌfleksəˈbɪləti]	n.	弹性
silicone rubber			硅橡胶
miscellaneous	[ˌmɪsəˈleɪniəs]	adj.	各种各样的

Lesson 7 Isobutylene-Based Elastomer (IIR)

Butyl rubber is the common name for the copolymer of isobutylene with 1% to 3% isoprene produced by cold (−100℃) cationic solution polymerization. The isoprene provides the unsaturation required for vulcanization. Most of butyl rubber's distinguishing characteristics are a result of its low level of chemical unsaturation. The essentially saturated hydrocarbon backbone of the IIR polymer will effectively repel water and polar liquids but show an affinity for aliphatic and some cyclic hydrocarbons. Products of butyl rubber will therefore be swollen by hydrocarbon solvents and oils, but show resistance to moisture, mineral acids, polar oxygenated solvents, synthetic hydraulic fluids, vegetable oils, and ester-type plasticizers. It is likewise highly resistant to the diffusion or solution of gas molecules. Air impermeability is the primary property of commercial utility. The low level of chemical unsaturation also imparts high resistance to ozone. [1] Sulfur-cured butyl rubber has relatively poor thermal stability, softening with prolonged exposure at temperatures above 150℃ because the low unsaturation prevents oxidative crosslinking. [2] Curing with phenol-formaldehyde resins instead of sulfur, however, provides products with very high heat resistance, the property responsible for a large market in tire-curing bladders.

阳离子溶液聚合
不饱和

饱和烃
极性液体

环烃

无机酸
酯类增塑剂

气密性

酚醛树脂

轮胎硫化胶囊

$$\left[-(CH_2-\underset{\underset{CH_3}{|}}{\overset{\overset{CH_3}{|}}{C}})_x-(CH_2-\underset{}{\overset{\overset{CH_3}{|}}{C}}=CH-CH_2)_y - \right]_n \quad y \ll x$$

IIR

[3] The molecular structure of the polyisobutylene chain provides less flexibility and greater delayed elastic response to deformation than most elastomers. This imparts vibration damping and shock-absorption properties to butyl rubber products.

振荡衰减/吸震作用

The unique properties of butyl rubber are used to advantage in tire inner tubes and air cushions (air impermeability), sheet roofing and cable insulation (ozone and weather resistance), tire-curing bladders, hoses for high temperature service, and conveyor belts for hot materials (thermal stability with resin cure).

绝缘电缆

传输带

New Words

cationic	[kætɪˈəʊnɪk]	adj.	阳离子的
unsaturation	[ˌʌnsætʃəˈreɪʃən]	n.	不饱和
characteristic	[ˌkærəktəˈrɪstɪk]	n.	特性
essentially	[ɪˈsenʃəli]	adv.	本质上,根本上

saturated	[ˈsætʃəreɪtɪd]	adj.	饱和的
backbone	[ˈbækbəʊn]	n.	主链
aliphatic	[ˌæləˈfætɪk]	adj.	脂肪族的
swell	[swel]	v.	溶胀
moisture	[ˈmɔɪstʃə(r)]	n.	水分；潮湿
hydraulic	[haɪˈdrɔːlɪk]	adj.	水力的，水压的
diffusion	[dɪˈfjuːʒn]	n.	扩散；传播
oxidative	[ˈɒksɪdeɪtɪv]	adj.	氧化的
impart	[ɪmˈpɑːt]	v.	给予
insulation	[ˌɪnsjuˈleɪʃn]	n.	绝缘

Notes

[1] Sulfur-cured butyl rubber has relatively poor thermal stability, softening with prolonged exposure at temperatures above 150℃ because the low unsaturation prevents oxidative crosslinking. 硫黄硫化的丁基橡胶热稳定性相对要差一些，若在超过150℃条件下长期使用会变软，这是因为它的低不饱和结构阻止了氧化交联。

[2] Curing with phenol-formaldehyde resins instead of sulfur, however, provides products with very high heat resistance, the property responsible for a large market in tire-curing bladders. 如果不用硫黄硫化，而用酚醛树脂硫化，丁基橡胶制品会得到更好的耐热性，这正是它能作为轮胎硫化胶囊而大范围使用的重要原因。

[3] The molecular structure of the polyisobutylene chain provides less flexibility and greater delayed elastic response to deformation than most elastomers. 聚异戊二烯结构中的大分子结构使丁基橡胶的弹性降低，而且其滞后现象比大多数弹性体要严重。

Exercises

1. Translate the first and second paragraphs of the text into Chinese.
2. Put the following words into English.

阳离子溶液聚合　丁基橡胶　　　饱和的　　　　溶胀　　　　轮胎硫化胶囊
气密性　　　　　极性的　　　　绝缘电缆

[Reading Material]

Halobutyl Rubbers

Halobutyl rubbers are produced by the controlled chlorination (chlorobutyl; CIIR) or bromination (bromobutyl; BIIR) of butyl rubber. Halogenation produces an allylic chlorine or bromine adjacent to the double bond of the isoprene unit. The halogenated butyl rubbers share many of the attributes of their butyl rubber parent: superior air impermeability, resistance to chemicals, moisture, and ozone, and vibration damping. The presence of halogen provides new crosslinking chemistry and the ability for adhesion to and vulcanization with general purpose, highly unsaturated elastomers. Bromobutyl is generally faster curing than chlorobutyl and somewhat more versatile in the curing systems that can be used to tailor product properties. Halobutyl rubbers typically use a zinc oxide-based cure, but bromobutyl can be vulcanized with peroxide or magnesium oxide as well.

The primary application for halobutyl rubber is in tires. The combination of low gas and moisture permeability, high heat and flex resistance, and ability to covulcanize with highly chemically unsaturated rubber has secured the use of these rubbers in the inner liners of tubeless tires. Passenger tires use chlorobutyl alone or in a blend with 20% to 40% natural rubber. High-service steel-belted truck tires use 100% bromobutyl inner liner compounds. Chlorobutyl is also used for truck inner tubes for its superior heat resistance compared to butyl rubber. Halobutyl rubbers are added to sidewall compounds for improved ozone and flex resistance, and to certain tread compounds for improved wet skid resistance and traction.

Halobutyl rubbers are also used in vibration damping mounts and pads, steam and automatic dishwasher hoses, chemical-resistant tank linings, and heat-resistant conveyor belts. They are also widely used in pharmaceutical closures because of their low gas and moisture permeability, resistance to aging and weathering, chemical and biological inertness, low extractables (with metal oxide cures), and good self-sealing and low fragmentation during needle penetration.

Words and Expressions

halobutyl rubber		n.	卤化丁基橡胶
chlorination	[ˌklɔːrɪ'neɪʃn]	n.	氯化
chlorobutyl	[klərəʊ'bɪtiːl]	n.	氯化丁基橡胶
bromination	[brəʊmɪ'neɪʃn]	n.	溴化
bromobutyl rubber		n.	溴化丁基橡胶
halogenation	[hælədʒə'neɪʃən]	n.	卤化
allylic	['æləˌlɪk]	adj.	烯丙基的
versatile	['vɜːsətaɪl]	adj.	多功能的
peroxide	[pə'rɒksaɪd]	n.	过氧化物
magnesium oxide			氧化镁
inner liner			气密层
tubeless	[t'juːbləs]	adj.	无内胎的
wet skid resistance			抗湿滑性
pharmaceutical	[ˌfɑːmə'suːtɪkl]	adj.	制药的
closure	['kləʊʒə(r)]	n.	封闭
biological inertness			生物惰性
extractable	[ɪks'træktəbl]	adj.	可抽出的,可萃取

Lesson 8 Ethylene Propylene Rubber (EPR)

The first commercial ethylene propylene rubbers were made by the random copolymerization of ethylene and propylene in solution using Ziegler-Natta catalysts. Since these compounds were fully saturated, they were highly resistant to oxidation, ozone, heat, weathering, and polar liquids. They could be cured, however, only by peroxide. [1] The greater versatility of a sulfur curable elastomer was sought, and found by the incorporation of limited amounts of a third monomer into the polymer. [2] Dienes were found which were compatible with the EPM polymerization process, and which had one double bond that would preferentially polymerize to leave a pendant double bond available for vulcanization. The latter criteria left the polymer backbone saturated and capable of offering the same high level of stability as EPM. The dienes most commonly used today to make the ethylene propylene diene monomers (EPDM) are 1,4-hexadiene, ethylidene norbornene and dicyclopentadiene, each conferring a different rate and state of cure to the polymer.

$$[(CH_2-CH_2)_3(CH(CH_3)-CH_2)(diene)_{0.2}]_n$$
EPDM

An extensive range of EPDM polymers are produced by varying the molecular weight, molecular weight distribution, ethylene/propylene ratio and level and type of diene termonomer. Elastomers are available containing from 50% to more than 75% ethylene by weight. Polymers with lower ethylene content are amorphous and easy to process. Higher ethylene content gives crystalline polymers with better physical properties, but more difficulty in processing. The amount of terpolymer is typically 1.5% to 4%, but can be as high as 11% in ultra-fast curing grades. Most EPDM are incompatible with diene rubbers (e.g., natural, SBR, NBR) because of their relatively slow cure rate. The ultra-fast cure EPDM overcome this.

In general, the ethylene propylene rubbers are compounded to provide good low-temperature flexibility, high tensile strength, high tear and abrasion resistance, excellent weatherability (ozone, water, oxidation resistance), good electrical properties, high compression set resistance, and high heat resistance. The high molecular weight crystalline EPDM can incorporate high levels of fillers. EPM and EPDM have low resistance to hydrocarbon oils and their lack of building tack must be compensated by the use of resin.

[3] The ethylene propylene rubbers are probably the most versatile of

the general purpose elastomers, and can be compounded in nearly the full spectrum of applications not requiring resistance to hydrocarbon oils. The high volume use is in tire sidewalls as an additive to improve ozone resistance. Other major uses which exploit their desirable properties and versatility are roof membrane and ditch liners, automotive seals, gaskets, weatherstripping and boots, appliance parts, and hosing. [4] Outside the rubber industry, they are used as viscosity modifiers in lubricating oils; they improve low temperature impact strength when added to polyolefins at low levels; and they form thermoplastic elastomers when blended in higher ratios with polyolefins or other thermoplastic resins.

防水卷材

黏度改性剂/润滑剂

热塑性弹性体

New Words

ethylene	[eθɪliːn]	n.	乙烯
propylene	[ˈprəʊpəliːn]	n.	丙烯
oxidation	[ˌɒksɪˈdeɪʃn]	n.	氧化
diene	[ˈdaɪiːn]	n.	二烯烃
preferential	[prefəˈrenʃl]	adj.	优先的
pendant	[ˈpendənt]	n.	垂饰
hexadiene	[ˈheksədɪən]	n.	己二烯
ethylidene	[əˈθɪlɪdiːn]	n.	亚乙基
norbornene	[nɔːrˈbɔːniːn]	n.	降冰片烯
dicyclopentadiene	[daɪsaɪkləpentəˈdriːn]	n.	二环戊二烯
extensive	[ɪkˈstensɪv]	adj.	范围广泛的
ratio	[ˈreɪʃiəʊ]	n.	比例
amorphous	[əˈmɔːrfəs]	adj.	非结晶的
terpolymer	[tɜːˈpɒlɪmə]	n.	三元共聚物
ultra-fast	[ˈʌltrəfˈɑːst]	adj.	超快
weatherability	[ˌweðrəˈbɪlɪtɪ]	n.	耐候性
spectrum	[ˈspektrəm]	n.	范围
modifier	[ˈmɒdɪfaɪə(r)]	n.	调节剂
lubricating	[ˈluːbrɪkeɪtɪŋ]	adj.	润滑的
polyolefin	[ˌpɒlɪˈəʊləfɪn]	n.	聚烯烃

Notes

[1] The greater versatility of a sulfur curable elastomer was sought, and found by the incorporation of limited amounts of a third monomer into the polymer. 可以通过引入少量的第三单体，使得乙丙橡胶能够用硫黄进行硫化，从而获得更全面的性能。

[2] Dienes were found which were compatible with the EPM polymerization process, and which had one double bond that would preferentially polymerize to leave a pendant double bond available for vulcanization. 研究发现二烯烃可以在乙丙橡胶聚合的过程中加入，这样可以使其中的一个双键优先参与聚合，而另一个双键作为侧基用于硫化反应。

[3] The ethylene propylene rubbers are probably the most versatile of the general purpose elastomers, and can be compounded in nearly the full spectrum of applications not requiring resistance to hydrocarbon oils. 乙丙橡胶几乎是性能最为全面的通用橡胶，它除了不耐有机油类，几乎可以满足所有的应用。

[4] Outside the rubber industry, they are used as viscosity modifiers in lubricating oils; they improve low temperature impact strength when added to polyolefins at low levels; and they form thermoplastic elastomers when blended in higher ratios with polyolefins or other thermoplastic resins. 除了橡胶工业外，它们还能作为润滑油的黏度改性剂使用；它们加入到聚烯烃类材料中可以提高低温抗冲击性能；它们和聚烯烃树脂或其他热塑性树脂共混，可以制备热塑性弹性体。

Exercises

1. Translate the first and second paragraphs of the text into Chinese.
2. Put the following words into English.

二元乙丙橡胶   三元乙丙橡胶   乙烯   丙烯   二烯烃
三元共聚物   导电性能   耐候性

[Reading Material]

Thermal Properties of EPDM

EPDM is largely unaffected by weather with very good resistance to ozone. DuPont literature quotes EPDM products which were exposed to 10000 parts per hundred million of ozone for 1000 hours in air, at room temperature, without cracking. Raw gum elastomer manufacturers' literature indicate upper continuous heat aging temperature limits in air, anywhere from 126℃ to around 150℃. Introducing time into this equation, one scenario might be: one cumulative month at 165℃, one cumulative year at 125℃, and five cumulative years at 100℃.

Low temperature flexibility is very good and compares well with NR, and like NR and SBR, EPDM (with a lower polarity than NR) has very poor oil resistance. The use of EPDM is dominant in roof membrane linings and extruded channels for windows. EPDM has also been used as a blend with NR in tire sidewalls to improve resistance to cracking by ozone attack. The excellent electrical resistance of EPDM promotes its use in medium and high voltage cable covers. Automotive applications of EPDM would include radiator and heater hoses and weather strips.

Words and Expressions

DuPont			美国杜邦公司
parts per hundred million (pphm)			亿分之几
heat aging			热老化
equation	[ɪˈkweɪʒn]	n.	反应式，方程式
scenario	[səˈnɑːriəʊ]	n.	方案
cumulative	[ˈkjuːmjələtɪv]	adj.	累积的

polarity	[pə'lærəti]	n.	极性
dominant	['dɒmɪnənt]	adj.	占优势的
roof membrane			防水卷材
radiator	[reɪdɪeɪtə(r)]	n.	散热器

Lesson 9　Nitrile Rubber (NBR)

　　Nitrile rubber is the generic name given to emulsion polymerized copolymers of acrylonitrile and butadiene. The production process itself is not overly complex; the polymerization, monomer recovery, and coagulation processes require some additives and equipment, but they are typical of the production of most rubbers. The necessary apparatus is simple and easy to obtain. For these reasons, the substance is widely produced in poorer countries where labor is relatively cheap. Among the highest producers of NBR are mainland China and Taiwan.

$$[(CH_2-CH=CH-CH_2)_2(\overset{\overset{C\equiv N}{|}}{CH}-CH_2)]_n$$
NBR

　　Its single most important property is exceptional resistance to attack by most oils and solvents. It also offers better air impermeability, abrasion resistance, and thermal stability than the general purpose elastomers like natural rubber and SBR. [1] These attributes arise from the highly polar character of acrylonitrile, the content of which determines the polymer's particular balance of properties. Commercial nitrile rubbers are available with acrylonitrile/butadiene ratios ranging from 18∶82 to 45∶55. [2] As acrylonitrile content increases, oil resistance, solvent resistance, tensile strength, hardness, abrasion resistance, heat resistance, and gas impermeability improve, but compression set resistance, resilience and low temperature flex deteriorate. Selection of the particular grade of NBR needed is generally based on oil resistance vs. low temperature performance. Blends of different grades are common to achieve the desired balance of properties.

　　[3] Nitrile elastomers do not crystallize when stretched and so require reinforcing fillers to develop optimum tensile strength, abrasion resistance, and tear resistance. They also possess poor building tack. [4] Although nitrile rubbers are broadly oil-resistant and solvent-resistant, they are susceptible to attack by certain strongly polar liquids, to which the nonpolar rubbers, such as SBR or natural rubber, are resistant. Nitrile rubber is poorly compatible with natural rubber, but can be blended in all proportions with SBR. This decreases overall oil resistance, but increases resistance to polar liquids in proportion to the SBR content. Nitrile polymers are increasingly used as additives to plastics to provide elastomeric properties. Blends with polyvinyl chloride are popular for conferring improved abrasion, tensile, tear, and flex properties.

　　The uses of nitrile rubber include, automotive transmission belts, hoses, O-rings, gaskets, oil seals, <u>v-belts</u>, synthetic leather, printer's form rollers, and as cable jacketing. NBR latex can also be used in the preparation of adhesives and as a <u>pigment binder</u>. A hydrogenated version of nitrile rubber, HNBR, also known as highly saturated nitrile (HSN), is commonly used to manufacture O-rings for automotive air-conditioning systems.

三角 V 带（三角带）	
颜料黏合剂	

New Words

acrylonitrile	[ˌækrələʊˈnaɪtrɪl]	n.	丙烯腈
exceptional	[ɪkˈsepʃənl]	adj.	杰出的
deteriorate	[dɪˈtɪəriəreɪt]	v.	使恶化
vs.(versus)	[ˈvɜːsəs]	prep.	对（比）
optimum	[ˈɒptɪməm]	adj.	最适宜的
susceptible	[səˈseptəbl]	adj.	易受影响的
proportion	[prəˈpɔːʃn]	n.	比，比率
hydrogenate	[ˈhaɪdrədʒəneɪt]	v.	使氢化

Notes

[1] These attributes arise from the highly polar character of acrylonitrile, the content of which determines the polymer's particular balance of properties. 得益于丙烯腈结构的高极性，丁腈橡胶表现出的综合性能非常好。

[2] As acrylonitrile content increases, oil resistance, solvent resistance, tensile strength, hardness, abrasion resistance, heat resistance, and gas impermeability improve, but compression set resistance, resilience and low temperature flex deteriorate. 随着丙烯腈含量的提高，丁腈橡胶的耐油性、耐溶剂性、拉伸强度、硬度、耐磨性、耐热性和气密性会提高，但是抗压缩永久变形性、弹性和低温回弹性会降低。

[3] Nitrile elastomers do not crystallize when stretched and so require reinforcing fillers to develop optimum tensile strength, abrasion resistance, and tear resistance. 丙烯腈类的弹性体拉伸时不会发生结晶现象，没有自补强性，因此需要加入补强填充剂以获得理想的拉伸强度、耐磨性和抗撕裂性能。

[4] Although nitrile rubbers are broadly oil-resistant and solvent-resistant, they are susceptible to attack by certain strongly polar liquids, to which the nonpolar rubbers, such as SBR or natural rubber, are resistant. 尽管丁腈橡胶被广泛应用于制备耐油和耐溶剂制品，但是它不像天然橡胶和丁苯橡胶这样的非极性橡胶一样具有耐强极性溶剂的性能。

Exercises

1. Translate the first and second paragraphs of the text into Chinese.
2. Put the following words into English.

丁腈橡胶	非极性橡胶	极性橡胶	三角 V 带	O 形圈
丙烯腈	耐油性	耐溶剂性		

[Reading Material]

Hydrogenated Nitrile Butadiene Rubber

Hydrogenated nitrile butadiene rubber (HNBR) is a relatively new elastomer, making its first appearance in 1984. The symbol for the generic material is HNBR, although HSN is sometimes used in literature, standing for highly saturated nitrile. It has all the attributes of NBR plus a very much higher heat resistance, dependent on the grade chosen. It also has very good weather and abrasion resistance, plus good mechanical strength. It is used in oilfields where it has resistance to amine corrosion inhibitors and better hydrogen sulfide resistance than NBR. It has established itself in automotive applications for timing belts, gaskets and O-rings, where higher temperature resistant elastomers are needed. Peroxide cured HNBR has heat aging resistance up to 150℃, based on around 1000 hours, while sulfur donor cured HNBR temperature resistance might drop to 135℃. Cost is somewhat less than conventional fluorocarbon rubber (FKM) on a weight basis, also since the density (using g/cm^3, which approximates to specific gravity) of HNBR is about half that of FKM, more products can be made for the same weight purchased.

Words and Expressions

oilfield	[ɔɪlfiːld]	n.	油田
amine	[əˈmiːn]	n.	胺
corrosion inhibitor			腐蚀抑制剂
sulfide	[ˈsʌlfaɪd]	n.	硫化物
timing belt			同步带
O-ring	[oʊ rɪŋ]		O形圈
approximate	[əˈprɒksɪmət]	adj.	大概的

Lesson 10   Neoprene (CR)

Neoprene is the common name for the polymers of chloroprene (2-chloro-1,3-butadiene). These are produced by emulsion polymerization. The chloroprene monomer can polymerize in four isomeric forms: *trans*-1,4-addition; *cis*-1,4-addition; 1,2-addition, leaving a pendant vinyl group and allylic chlorine; and 3,4-addition. Neoprene is typically 88%~92% trans, with degree of polymer crystallinity proportional to the trans content. Cis addition accounts for 7%~12% of the structure and 3,4-addition makes up about 1%. The approximately 1.5% of 1,2-addition is believed to provide the principal sites of vulcanization.

$$\left[CH_2-\underset{\underset{CR}{|}}{\overset{\overset{Cl}{|}}{C}}=CH-CH_2\right]_n$$

[1] The high structural regularity (high-trans content) of neoprene allows the strain-induced crystallization that results, as for natural rubber, in high tensile strength. [2] The 2-chloro substituent, instead of natural rubber's 2-methyl, results in a higher freezing point (poorer low temperature resistance) and alters vulcanization requirements. Neoprenes are generally cured with zinc oxide and magnesium oxide, or lead oxide for enhanced water resistance. The presence of chlorine in the polymer structure improves resistance to oil, weathering, ozone and heat. The improved oxidation resistance is due to the reduced activity of the double bonds caused by the chlorine. Except for low temperature resistance and price, neoprene would be considered nearly as versatile as natural rubber.

[3] There are three types of general purpose neoprenes—G, W, and T types—with selected features modified to offer a range of processing, curing and performance properties. Products are made from neoprene because it offers good building tack, good oil, abrasion, chemical, heat, weather, and flex resistance, and physical toughness. Neoprene is widely used in hoses of all types (water, oil, air, automotive, industrial), wire and cable jacketing, power transmission and conveyor belting, bridge and building bearings, pipe gaskets, footwear, roof coatings, and coated fabrics.

氯丁橡胶/2-氯-1,3-丁二烯

顺式加成

拉伸诱导结晶

耐低温性能

氧化铅

抗氧化性

电力传输/传送带

New Words

neoprene	[ˈniːəpriːn]	*n.*	氯丁(二烯)橡胶
chloroprene	[ˈklɔːrəpriːn]	*n.*	氯丁二烯
principal	[ˈprɪnsəpl]	*adj.*	主要的
substituent	[sʌbˈstɪtʃuənt]	*n.*	取代
activity	[ækˈtɪvəti]	*n.*	活性

Notes

[1] The high structural regularity (high-trans content) of neoprene allows the strain-induced crystallization that results, as for natural rubber, in high tensile strength. 反式结构含量高的氯丁橡胶具有较高的结构规整性,像天然橡胶一样,这会使氯丁橡胶产生拉伸诱导结晶现象,从而获得较高的拉伸强度。

[2] The 2-chloro substituent, instead of natural rubber's 2-methyl, results in a higher freezing point (poorer low temperature resistance) and alters vulcanization requirements. 2-氯取代基代替天然橡胶的 2-甲基,导致氯丁橡胶具有更高的熔点(耐低温性变差),并改变了硫化的要求。

[3] There are three types of general purpose neoprenes—G, W, and T types—with selected features modified to offer a range of processing, curing and performance properties. 氯丁橡胶具有三种类型:G 型、W 型和 T 型,它们各自在加工、硫化和性能上具有不同的特点。

Exercises

1. Translate the first and second paragraphs of the text into Chinese.
2. Put the following words into English.

氯丁橡胶　　　顺式加成　　　氯丁二烯　　　取代　　　双键活性
传送带

[Reading Material]

The Life of Neoprene

The CR in the heading stands for chloroprene rubber, more popularly known as Neoprene. Like all of the synthetic elastomers, CR is available to the rubber chemist in a number of grades to aid in compound mixing and to emphasize certain properties, such as reduction of crystallization rate in the vulcanizate. CR has a measure of both oil and weather resistance. The oil resistance would only be considered moderate. CR has similar dynamic mechanical characteristics to NR, including good mechanical strength when it is compounded as a gum vulcanizate. CR has some ability to retard flame, which means that when a source of flame is removed, the burning polymer will have a tendency to self extinguish, while NR, EPDM, and SBR for example, will continue to burn. Upper continuous heat aging resistance temperature limits are of the order of 90℃. Like a number of elastomers this can be raised somewhat by special compounding.

It is common to see the word continuous used in the literature without reference to time of exposure (is it a month, a year, or ten years?) and also there is no definition of what constitutes ultimate failure of the material. An American standard SAE J2236 defines continuous upper temperature resistance as, the temperature at which the material retains a minimum of 50% of both original elongation and tensile strength at break after 1008 hours (6weeks). This will be a good precision reference for engineers and chemists as laboratories submit their materials to this standard. Leaving this kind of precision behind, one suggested upper range for CR in air is 99℃ for 1000 cumulative hours and 85℃ for 10000 cumulative hours. At the opposite end of the temperature range, CR shows some stiffening at around $-18°C$, becoming brittle around $-40°C$, although this can be lowered using certain compounding ingredients. Resistance of CR to dilute acids and bases is better than that of NR or SBR, while cost is somewhat higher. One last point, certain grades of CR are produced specifically for the adhesives marketplace.

Words and Expressions

emphasize	['emfəsaɪz]	v.	强调, 着重
reduction	[rɪ'dʌkʃn]	n.	减少
crystallization rate			结晶速率
moderate	['mɒdərət]	adj.	适度的, 中等的
dynamic mechanical characteristics			动态力学性能
retard	[rɪ'tɑːrd]	v.	阻止, 推迟
self extinguish			自熄
minimum	['mɪnəməm]	n.	最小量
stiffen	['stɪfən]	v.	(使)变硬
brittle	['brɪtl]	adj.	易碎的
ingredient	[ɪn'griːdiənt]	n.	原料
dilute acid			稀酸

Lesson 11 Chlorinated and Chlorosulfonated Polyethylene (CM, CSM)

Chlorinated polyethylene (CM) is produced by the chlorination of high density polyethylene either in solvent solution or aqueous suspension. The substituent chlorine on the saturated olefin backbone enhances heat and oil resistance. The chlorine also provides flame resistance. The polymer is thermoplastic when processed on conventional elastomer equipment, and compounds can be molded, calendered or extruded. Chlorinated polyethylene is most often steam cured using a peroxide curing system. The major end use is wire and cable applications, particularly flexible cords for up to 600 volts. Other major uses are in automotive hose, sheet goods and as an impact modifier in plastics.

$$\left[CH_2-CH_2-\underset{\underset{Cl}{|}}{CH}-CH_2-CH_2 \right]_n$$
CM

$$\left[(CH_2-CH_2-CH_2-\underset{\underset{Cl}{|}}{CH}-CH_2-CH_2-CH_2)_{12}(\underset{\underset{SO_2Cl}{|}}{CH}-CH_2) \right]_n$$
CSM

Simultaneous chlorination and chlorosulfonation of high density polyethylene in an inert solvent yields chlorosulfonated polyethylene (CSM). Various grades are available with chlorine contents ranging from 24% to 43% and sulfur from 1.0% to 1.4%. [1] CSM is widely used in the rubber industry because its compounds are odorless, light stable, and easily colored, highly resistant to oxygen, ozone, weathering, and corrosive chemicals, and resistant to abrasion, heat, low temperatures, oil, and grease. They also have excellent electrical properties and provide high tensile properties without highly reinforcing fillers. The particular degree and balance of these attributes is governed primarily by chlorine content. Increasing chlorine gives increasing flame and oil resistance, decreasing heat and electrical resistance and low temperature flexibility, and slightly decreasing ozone resistance.

Uncured chlorosulfonated polyethylene is more thermoplastic than other commonly used elastomers. It is generally tougher at room temperature, but softens more rapidly on heating. Vulcanization can be obtained with peroxides, as with other chemically saturated polymers, or with sulfur crosslinking at the sulfonyl chloride groups. The former promotes better resistance to heat and compression set, the latter can result in high tensile strength and excellent mechanical toughness. Compounds can also be cured at ambient temperatures by the combination of water (atmospheric moisture) and a divalent metal oxide (magnesia), which form sulfonate salt bridges. These ionic cures are slow but develop high modulus.

Ambient temperature ionic curing of CSM has led to its major use in single-ply roofing membranes, and geomembranes for reservoir liners and waste dump liners. Roofing membranes are installed in an uncured thermoplastic state, with seaming by heat or solvents. Weathering slowly cures the membrane, progressively increasing durability and toughness. Because these compounds are easily colored, they can be made with heat reflective or absorptive colors as required. An alternative approach to weather resistant roofing is to coat a neoprene or EPDM base with CM paint. [2] The combination of colorability, toughness, environmental durability and resistance to flame, oil, radiation and corrosive chemicals has also secured CSM's widespread use in automotive hoses, tubes, electrical wire insulation for up to 600 volts, wire insulation in nuclear power stations, industrial hoses and tank linings, coated fabrics, hot conveyor belting, and construction coatings and gaskets.

大气湿度
二价金属氧化物
盐桥

防水卷材

核电站
储（油）罐衬里

New Words

aqueous	[eɪkwiəs]	adj.	水的
suspension	[sə'spenʃn]	n.	悬浮液
olefin	[əʊləfɪn]	n.	石蜡
volt	[vəʊlt]	n.	伏特
simultaneous	[ˌsɪml'teɪniəs]	adj.	同时的
chlorination	[ˌklɔːrɪ'neɪʃn]	n.	氯化
chlorosulfonation	[kləʊrəsʌlfə'neɪʃən]	n.	氯磺化
odorless	[əʊdələs]	adj.	无臭的
corrosive	[kə'rəʊsɪv]	adj.	腐蚀性的
grease	[griːs]	n.	润滑油
geomembrane	[dʒiːəʊmemb'reɪn]	n.	土工膜
reservoir	['rezəvwɑː(r)]	n.	蓄水池
seam	[siːm]	v.	接合
alternative	[ɔːl'tɜːnətɪv]	adj.	替代的
colorability	[ˌkʌlərə'bɪlɪti]	n.	可着色性

Notes

[1] CSM is widely used in the rubber industry because its compounds are odorless, light stable, and easily colored, highly resistant to oxygen, ozone, weathering, and corrosive chemicals, and resistant to abrasion, heat, low temperatures, oil, and grease. 氯磺化聚乙烯橡胶具有无臭、光稳定、易着色、耐氧和臭氧性好、耐候性好、耐化学腐蚀、耐磨、耐热、耐低温、耐油和润滑剂等优点，因此在橡胶工业中有广泛的应用。

[2] The combination of colorability, toughness, environmental durability and resistance to flame, oil, radiation and corrosive chemicals has also secured CSM's widespread use in automotive hoses, tubes, electrical wire insulation for up to 600 volts, wire insulation in nuclear power stations, industrial hoses and tank linings, coated fabrics, hot conveyor belting, and construction coatings and gaskets. 氯磺化聚乙烯橡胶具有易着色、强度高、耐久性好、阻燃、耐油、耐辐射、耐腐蚀等优点，因此它广泛应用在汽车胶管、600V 的绝缘电线、核电站的绝缘线、工业胶管、储罐衬里、织物图层、高温传送带、建筑涂料和减振垫等诸多领域。

Exercises

1. Translate the first and second paragraphs of the text into Chinese.
2. Put the following words into English.

| 悬浮液 | 石蜡 | 氯化聚乙烯 | 氯磺化聚乙烯 | 无臭的 |
| 氯化 | 可着色性 | | | |

[Reading Material]

Application of CM and CSM

Chlorinated polyethylene is an inexpensive variation of polyethylene having chlorine content from 34% to 44%. It is used in blends with polyvinyl chloride (PVC) because the soft, rubbery chlorinated polyethylene is embedded in the PVC matrix, thereby increasing the impact resistance. In addition, it also increases the weather resistance. Furthermore, it is used for softening PVC foils, without risking plasticizer migration. Chlorinated polyethylene can be crosslinked peroxidically to form an elastomer which is used in cable and rubber industry. When chlorinated polyethylene is added to other polyolefins, it reduces the flammability. Chlorinated polyethylene is sometimes used in power cords as an outer jacket.

In polymer science, chlorosulfonated polyethylene (CSM) is noted for its resistance to chemicals, temperature extremes, and UV light. Along with PVC, CSM is one of the most common materials used to make inflatable boats and folding kayaks. It is also used in roofing materials and as a surface coat material on radomes owing to its radar-transparent quality. CSM is also used in the construction of the decking of modern snowshoes, replacing neoprene as a lighter, stronger alternative.

Words and Expressions

rubbery	['rʌbəri]	adj.	有弹性的
embed	[ɪm'bed]	v.	把…嵌入
matrix	['meɪtrɪks]	n.	基质
foil	[fɔɪl]	n.	薄片，衬底
plasticizer	['plæstɪsaɪzə]	n.	增塑剂
migration	[maɪ'greʃn]	n.	迁移
polyolefin	[ˌpɒlɪ'əʊlefɪn]	n.	聚烯烃

flammability	[ˌflæməˈbɪlɪtɪ]	n.	可燃性
UV light			紫外线
radome	[ˈreɪdəʊm]	n.	天线罩

Lesson 12 Silicone Rubber (MQ, MVQ, MPVQ)

[1] Because of its unique properties and somewhat higher price compared to the other common elastomers, silicone rubber is usually classed as a specialty elastomer, although it is increasingly used as a cost-effective alternative in a variety of applications. Two types of silicone elastomers are available, each providing the same fundamental properties. These are the thermosetting rubbers that are vulcanized with heat, and RTV (room temperature vulcanizing) rubbers.

硅橡胶
特种橡胶

室温硫化橡胶

The basic silicone polymer is dimethylpolysiloxane with a backbone of silicon-oxygen linkages and two methyl groups on each silicon. [2] The silicon-oxygen backbone provides a high degree of inertness to ozone, oxygen, heat (up to 315℃), UV light, moisture, and general weathering effects, while the methyl substituents confer a high degree of flexibility. The basic polymer properties are modified by replacing minor amounts of the methyl substituents with phenyls and/or vinyls. Phenyl groups improve low temperature flexibility (to as low as －100℃) without sacrificing high temperature properties. Vinyl groups improve compression set resistance and facilitate vulcanization. Of the available silicone elastomers—methyl silicone (MQ), methyl-vinyl silicone (MVQ), methyl-phenyl silicone (MPQ), methyl-phenyl-vinyl silicone (MPVQ), and fluoro-vinyl-methyl silicone (FVMQ)—the methyl-vinyl types are most widely used.

硅氧主链

苯基基团

乙烯基基团

甲基硅橡胶/甲基乙烯基硅橡胶
氟硅胶

Thermal vulcanization typically uses peroxides to crosslink at the vinyl groups of the high molecular weight solid silicone rubbers. [3] Compounded products offer the attributes noted above plus superior resistance to compression set, excellent biocompatibility, vibration damping over a wide temperature range, and thermal ablative properties. Silicone elastomers generally offer poorer tensile, tear, and abrasion properties than the more common organic rubbers, but this is routinely improved by reinforcement with fumed silica, which also improves electrical insulation properties.

热硫化

气相法白炭黑/电绝缘性能

Room temperature vulcanizing (RTV) silicones are low molecular weight dimethylpolysiloxane liquids with reactive end groups. [4] As with the heat cured polymers, there can be minor substitution of methyl groups with phenyls—for improved low temperature flexibility, or with fluoroalkyl groups—for improved oil and solvent resistance and even broader temperature service. Vulcanization of the RTV silicones is obtained from either a condensation or an addition reaction.

可反应性端基

加成反应

Most fabricators of silicone rubber products do not do their own compounding, but purchase premixed compounds requiring only catalyst and/

or curing. Solid (thermally cured) rubbers are used in automotive under-hood applications, primarily for their heat resistance. Products include ignition cables, coolant and heater hoses, O-rings, and seals. Similar applications are found in aircraft seals, connectors, cushions, and hoses, and in home appliance O-rings, seals, and gaskets. Long service life plus circuit integrity and no toxic gas generation have secured the place of silicone rubber in wire and cable insulation for electric power generation and transmission, for naval shipboard cable of all types, and for appliance wiring. The inherent inertness and biocompatibility of silicone rubbers have enabled their use in food contact and medical products. These include baby bottle nipples, belts and hoses for conveying foods and food ingredients, surgical tubing, and prosthetic devices. RTV silicones are used by the automotive, appliance, and aerospace industries for electronic potting compounds and formed-in-place gaskets, to form molds for the manufacture of plastic parts, and widely in construction adhesives, sealants, roof coatings, and glazing.

家用电器
有毒气体
食品添加剂

New Words

silicone	['sɪlɪkəʊn]	n.	硅树脂
dimethylpolysiloxane	[daɪmeθɪlpɒliːˈsaɪlɒksæn]	n.	二甲聚硅氧烷
methyl	['meθɪl]	n.	甲基
inertness	[ɪnˈɜːtnɪs]	n.	惰性
phenyl	['fenəl]	n.	苯基
vinyl	['vaɪnl]	n.	乙烯基
biocompatibility	[biːəʊkəmpætəˈbɪlɪtɪ]	n.	生物相容性
ablative	['æblətɪv]	n.	离格
fluoroalkyl	['fluːəˈælkɪl]	n.	氟烷基
condensation	[kɒndenˈseɪʃn]	n.	缩合聚合
alkoxy	['ɔːlkɒksɪ]	adj.	烷氧基的
premix	[priːˈmɪks]	adj.	预混
underhood	['ʌndərhʊd]	n.	发动机舱
ignition	[ɪgˈnɪʃn]	n.	（引擎的）点火装置
coolant	[kuːlənt]	n.	冷冻剂
integrity	[ɪnˈtegrəti]	n.	完整
surgical	['sɜːdʒɪkl]	adj.	外科手术的
prosthetic	[prɒsˈθetɪk]	adj.	义肢的，假体的
aerospace	[ˈeərəʊspeɪs]	n.	航空与航天

Notes

[1] Because of its unique properties and somewhat higher price compared to the other common elastomers, silicone rubber is usually classed as a specialty elastomer, although it is increas-

ingly used as a cost-effective alternative in a variety of applications. 硅橡胶因其独一无二的性能和相对通用橡胶较为昂贵的价格，通常被划分为特种橡胶，但是性价比高让它在很多领域应用越来越广泛。

[2] The silicon-oxygen backbone provides a high degree of inertness to ozone, oxygen, heat (up to 315℃), UV light, moisture, and general weathering effects, while the methyl substituents confer a high degree of flexibility. 硅氧主链结构让硅橡胶在臭氧、氧、热（高达315℃）、紫外线、高湿度、天候等作用下，表现出优异的化学惰性，同时硅上的甲基赋予了硅橡胶高弹性。

[3] Compounded products offer the attributes noted above plus superior resistance to compression set, excellent biocompatibility, vibration damping over a wide temperature range, and thermal ablative properties. 配合好的硅橡胶产品除了具有上面提到的性能外，还具有更好的抗压缩永久变形性、优异的生物相容性、较宽温度范围内的小滞后损耗和隔热性能。

[4] As with the heat cured polymers, there can be minor substitution of methyl groups with phenyls—for improved low temperature flexibility, or with fluoroalkyl groups—for improved oil and solvent resistance and even broader temperature service. （热硫化性硅橡胶）作为热硫化聚合物，用少量苯基取代甲基以提高低温弹性，也可以用少量氟烷基取代甲基以获得耐油性、耐溶剂性和更宽的温度使用范围。

Exercises

1. Translate the first and second paragraphs of the text into Chinese.
2. Put the following words into English.

硅橡胶　　　　甲基硅橡胶　　　甲基乙烯基硅橡胶　　　甲基乙烯基苯基硅橡胶　　氟硅胶
热硫化性硅橡胶　室温硫化硅橡胶　生物相容性

[Reading Material]

RTV Silicones

Vulcanization of the RTV silicones is obtained from either a condensation or an addition reaction. Condensation cures can be either moisture independent or moisture dependent. For moisture independent compounds, the reactive polymer end group is usually silanol. The crosslinking agent may be a silicone with silanol end groups, using an organic base as the condensation catalyst, an alkoxy silicate (e.g., ethyl silicate), using a metallic salt catalyst, or a polyfunctional silicone, often requiring no catalyst. These compounds are known as two-package RTV since the curing agent and/or catalyst is kept separate and added to the compound just prior to use. Moisture-curing compounds, also known as one-package RTV, are compounded from silanol-terminated polymer with a polyfunctional silane curing agent (e.g., methyltriacetoxysilane) and condensation catalyst, or from a polymer end-stopped with the curing agent. Crosslinking occurs on exposure to atmospheric moisture, starting at the surface and progressing inward with diffusion of moisture into the compound.

Addition-cured RTV are typically compounded from a dimethylvinylsiloxy-terminated polymer, a polyfunctional silicon hydride crosslinker, and a metal ion catalyst. Vulcanization is independent of moisture and air and forms no volatile byproducts. These products are usually sold as two-package RTV, but are also available as one-package compounds containing an inhibitor which is volatilized or deactivated by heat to trigger the cure.

Words and Expressions

silanol	[ˈsɪleɪnɒl]	n.	硅烷醇
alkoxy silicate			烷氧基硅
polyfunctional	[ˌpɒlɪˈfʌŋkʃənəl]	adj.	多官能的
two-package			双组分
one-package			单组分
moisture-curing			湿气固化
silicon hydride			硅烷
metal ion catalyst			金属离子催化剂
volatilize	[ˈvɒlətɪlaɪz]	v.	挥发
deactivate	[ˌdiːˈæktɪveɪt]	v.	使无效

Lesson 13 Fluorocarbon Rubber (FKM)

In the United States <u>fluorocarbon rubber</u> is well known by its trade name of Viton. Based on <u>vinylidene fluoride</u> and <u>hexafluoropropylene</u> the grades available differ in the chemical building blocks which were used to construct the polymer. Like silicone rubber, FKM has excellent high temperature resistance with an upper continuous <u>heat aging</u> temperature limit of 205℃. <u>DuPont</u> literature quotes continuous dry heat service to be＞3000 hours at 232℃ decreasing to＞48 hours at 316℃.

At the opposite end of the scale the conventional FKM is usually serviceable at temperatures down to －20℃ in dynamic applications, while for static use the temperature can be lower, although this will depend on the grade chosen. A primary variable in FKM grades is the level of fluorine in the elastomer molecule, FKM being fluorohydrocarbons. Terpolymers tend to have a higher fluorine content than copolymers and therefore have better resistance to various media. In general, fluoroelastomers have excellent resistance to oxidation, ozone, fuels and petroleum oils and are resistant to most <u>mineral acids at high concentrations</u>. Although FKM has good resistance to many chemicals, excessive swelling occurs in some polar solvents such as low molecular weight ethers, esters and ketones. [1] Chemicals such as alkalis and amines should be used with caution, with standard fluorocarbon grades, especially at higher temperatures because alkalis harden the general purpose FKM, which will eventually embrittle and then crack.

FKM has a tendency to <u>self extinguish</u> when a flame is removed. This is of benefit in situations where the results of a fire would be catastrophic, for example in a coalmine. Other elastomers might burn out of control, when the source of the originating flame (such as methane <u>gas explosion</u>) is removed. [2] Applications for FKM include <u>automotive fuel hose</u> liners and seals and flue duct expansion joints, where high temperatures and acidic products from gas desulfurization are involved. [3] The relative cost of FKM is high, more than any of the elastomers mentioned so far, also a high specific gravity (around 1.8) means less cured product (volume) per unit weight.

New Words

fluorocarbon	[ˌfluːərəʊˈkɑːbən]	n.	碳氟化合物
vinylidene	[vaɪˈnɪlɪdiːn]	n.	亚乙烯基
fluoride	[ˈflɔːraɪd]	n.	氟化物
hexafluoropropylene	[heksɑːfˈluːɔːrəʊprɒpiːliːn]	n.	六氟丙烯

serviceable	['sɜːvɪsəbl]	adj.	耐用的
static	['stætɪk]	adj.	静止的
fluorine	['flɔːriːn]	n.	氟
concentration	[kɒnsn'treɪʃn]	n.	浓度
ether	[iːθə(r)]	n.	醚
ester	['estə(r)]	n.	酯
ketone	['kiːtəʊn]	n.	酮
alkali	[ælkəlaɪ]	n.	碱
amine	[ə'miːn]	n.	胺
embrittle	[ɪm'brɪtəl]	v.	使变脆
catastrophic	[kækə'strɒfɪk]	adj.	灾难的
coalmine	[kəʊl'maɪn]	n.	煤矿
methane	[miːθeɪn]	n.	甲烷
desulfurization	[diːsʌlfəraɪ'zeɪʃn]	n.	脱硫

Notes

[1] Chemicals such as alkalis and amines should be used with caution, with standard fluorocarbon grades, especially at higher temperatures because alkalis harden the general purpose FKM, which will eventually embrittle and then crack. 一般的氟橡胶在使用时应避免与类似碱和胺的化学品接触，尤其在高温下，这是因为碱能够使氟橡胶发生硬化变脆，进而脆裂。

[2] Applications for FKM include automotive fuel hose liners and seals and flue duct expansion joints, where high temperatures and acidic products from gas desulfurization are involved. 氟橡胶主要用来制备汽车油管内衬、油封、排气管接头等，多是在接触来自于气体脱硫的高温、酸产品的地方。

[3] The relative cost of FKM is high, more than any of the elastomers mentioned so far, also a high specific gravity (around 1.8) means less cured product (volume) per unit weight. 氟橡胶与其他橡胶相比，一方面价格要昂贵，另一方面它的相对密度较大（1.8左右），相同质量的橡胶，只能获得相对较少的硫化胶产品。

Exercises

1. Translate the first and second paragraphs of the text into Chinese.
2. Put the following words into English.

氟橡胶	碳氟化合物	氟化物	静止的	浓度
醚	甲烷	氟	酯	

[Reading Material]

Fluoroelastomers

Fluoroelastomers—these are chemically saturated copolymers and terpolymers of vinylidene fluoride, hexafluoropropylene, tetrafluoroethylene, perfluoro (methyl vinyl) ether, and propylene in various combinations. They are designed to provide extraordinary levels of resistance to oil, chemicals, and heat. They are generally classified into four groups: A, B, F, and Specialty. The lettered groups have increasing fluid resistance, reflecting their increasing flourine levels of 66%, 68%, and 70% respectively. The Specialty group contains further enhanced properties, such as improved low temperature flexibility. The rest are most often cured with bisphenol. Because of their exceptional resistance to heat aging and a broad range of chemicals, fuels and solvents, fluoroelastomers are used in a wide variety of demanding automotive, aerospace, and industrial applications. These include seals, gaskets, liners, hoses, protective fabric coatings, diaphragms, roll covers, and cable jacketing.

Words and Expressions

fluoroelastomer	[fluːərərˈlɑːstəmə]	n.	含氟弹性体
tetrafluoroethylene	[tetrəˈfluərəˈeθiliːn]	n.	四氟乙烯
perfluoro-			全氟代
bisphenol	[ˈbɪsfɪnɒl]	n.	双酚

Lesson 14　Thermoplastic Elastomers (TPE)

　　Thermoplastic elastomers (TPE), sometimes referred to as thermoplastic rubbers, are a class of copolymers or a physical mix of polymers (usually a plastic and a rubber) which consist of materials with both thermoplastic and elastomeric properties. While most elastomers are thermosets, thermoplastics are in contrast relatively easy to use in manufacturing, for example, by injection molding. Thermoplastic elastomers show advantages typical of both rubbery materials and plastic materials. [1] The benefit of using thermoplastic elastomers is the ability to stretch to moderate elongations and return to its near original shape creating a longer life and better physical range than other materials. The principal difference between thermoset elastomers and thermoplastic elastomers is the type of crosslinking bond in their structures. In fact, crosslinking is a critical structural factor which imparts high elastic properties.

　　[2] TPE materials have the potential to be recyclable since they can be molded, extruded and reused like plastics, but they have typical elastic properties of rubbers which are not recyclable owing to their thermosetting characteristics. They can also be ground up and turned into 3D printing filament. TPE also require little or no compounding, with no need to add reinforcing agents, stabilizers or cure systems. [3] Hence, batch-to-batch variations in weighting and metering components are absent, leading to improved consistency in both raw materials and fabricated articles. Depending on the environment, TPE have outstanding thermal properties and material stability when exposed to a broad range of temperatures and non-polar materials. TPE consume less energy to produce, can be colored easily by most dyes, and allow economical quality control.

　　The two most important manufacturing methods with TPE are extrusion and injection molding. TPE can now be 3D printed and have been shown to be economically advantageous to make products using distributed manufacturing. Compression molding is seldom, if ever, used. Fabrication via injection molding is extremely rapid and highly economical. Both the equipment and methods normally used for the extrusion or injection molding of a conventional thermoplastic are generally suitable for TPE. TPE can also be processed by blow molding, melt calendaring, thermoforming, and heat welding.

　　TPE are used where conventional elastomers cannot provide the range of physical properties needed in the product. These materials find large application in the automotive sector and in household appliances sector. About 40% of all TPE products are used in the manufacturing of vehicles.

For instance <u>copolyester</u> TPE are used in snowmobile tracks where stiffness and abrasion resistance are at a premium. <u>Thermoplastic olefins (TPO)</u> are increasingly used as a roofing material. TPE are also widely used for catheters where nylon block copolymers offer a range of softness ideal for patients. Thermoplastic silicone and olefin blends are used for extrusion of glass run and dynamic weatherstripping car profiles. Styrene block copolymers are used in shoe soles for their ease of processing, and widely as adhesives. [4] TPE is commonly used to make suspension bushings for automotive performance applications because of its greater resistance to deformation when compared to regular rubber bushings. Thermoplastics have experienced growth in the heating, ventilation, and air conditioning industry due to the function, cost effectiveness and adaptability to modify plastic resins into a variety of covers, fans and housings. TPE may also be used in medical devices and is also finding more and more uses as an electrical cable jacket and inner insulation. You'll also be able to find TPE used in some headphone cables.

共聚酯
热塑性聚烯烃

New Words

manufacture	[mænjuˈfæktʃə(r)]	v.	制造
thermoset	[ˈθɜːməset]	adj.	热固性的
injection	[ɪnˈdʒekʃn]	n.	注射
recyclable	[ˌriːˈsaɪkləbl]	adj.	可循环再用的
filament	[ˈfɪləmənt]	n.	单纤维
dye	[daɪ]	n.	染料
economical	[ˌiːkəˈnɒmɪkl]	adj.	经济的
thermoform	[ˈθɜːməˌfɔːm]	v.	热成型
weld	[weld]	v.	焊接
catheter	[ˈkæθɪtə(r)]	n.	导尿管
nylon	[ˈnaɪlɒn]	n.	尼龙
weatherstrip	[ˈweðəstrɪp]	n.	挡风雨条
headphone	[ˈhedfəʊn]	n.	双耳式耳机

Notes

[1] The benefit of using thermoplastic elastomers is the ability to stretch to moderate elongations and return to its near original shape creating a longer life and better physical range than other materials. 热塑性弹性体的主要优点是拉伸后具有适度的伸长率，恢复后能基本达到原来的形状，这赋予了它更长的使用寿命和更好的物理机械性能。

[2] TPE materials have the potential to be recyclable since they can be molded, extruded and reused like plastics, but they have typical elastic properties of rubbers which are not recyclable owing to their thermosetting characteristics. 热塑性弹性体具有很好的可回收再用前景，因为它可以像塑料一样模塑、挤出和再加工，又具有橡胶一样的弹性，然而橡胶通常是热固性的不能再

回收。

[3] Hence, batch-to-batch variations in weighting and metering components are absent, leading to improved consistency in both raw materials and fabricated articles. 因此，各批次间由称量和计量所带来的误差消除了，进而提高了原材料和制品的一致性。

[4] TPE is commonly used to make suspension bushings for automotive performance applications because of its greater resistance to deformation when compared to regular rubber bushings. 与普通橡胶相比，热塑性弹性体通常用于制造用于汽车性能应用的悬架衬套，因为它具有更大的抗变形能力。

Exercises

1. Translate the first and second paragraphs of the text into Chinese.
2. Put the following words into English.

热塑性弹性体　　热固性的　　可循环再用的　　模压　　注射
吹塑成型　　热硫化性弹性体

[Reading Material]

Nanostructure of TPE

It was not until the 1950s, when thermoplastic polyurethane polymers became available, that TPE became a commercial reality. During the 1960s styrene block copolymer became available, and in the 1970s a wide range of TPE came on the scene. The worldwide usage of TPE is growing at about 9% per year. The styrene-butadiene materials possess a two-phase microstructure due to incompatibility between the polystyrene and polybutadiene blocks, the former separating into spheres or rods depending on the exact composition. With low polystyrene content, the material is elastomeric with the properties of the polybutadiene predominating. Generally, they offer a much wider range of properties than conventional cross-linked rubbers because the composition can vary to suit final construction goals.

Block copolymers are interesting because they can "microphase separate" to form periodic nanostructures, as in the styrene-butadiene-styrene (SBS) block copolymer. The polymer is used for shoe soles and adhesives. The material was made by living polymerization so that the blocks are almost monodisperse, so helping to create a very regular microstructure. Since most polymers are incompatible with one another, forming a block polymer will usually result in phase separation, and the principle has been widely exploited since the introduction of the SBS block polymers, especially where one of the block is highly crystalline. One exception to the rule of incompatibility is the material noryl, where polystyrene and polyphenylene oxide (PPO) form a continuous blend with one another.

Other TPEs have crystalline domains where one kind of block co-crystallizes with other block in adjacent chains, such as in copolyester rubbers, achieving the same effect as in the SBS block polymers. Depending on the block length, the domains are generally more stable than the latter owing to the higher crystal melting point. That point determines the processing temperatures needed to shape the material, as well as the ultimate service use temperatures of the product.

Words and Expressions

polyurethane	[ˌpɒliˈjuərəθeɪn]	n.	聚氨酯
two-phase microstructure			两相微观结构
incompatibility	[ˌɪnkəmˌpætəˈbɪləti]	n.	不相容
microphase separate			微观相分离
periodic	[pɪəriˈɒdɪk]	adj.	周期的
nanostructure	[næˈnɒstrʌktʃər]	n.	纳米结构
monodisperse	[mɒnoʊdɪsˌpɜːs]	adj.	单分散(性)的

PART 3

Introduction to Rubber Compounding

Lesson 15 Introduction to Rubber Compounding

[1] Rubber compounding includes selection of proper ingredients to enable one to process compound and vulcanize to give desirable physical and chemical properties, especially after aging. Elastomers by themselves are not useful until they are properly mixed with the proper ingredients and cured. Depending on the type of rubber products, various ingredients are used. Each ingredient has specific function in imparting desired properties.

Compounding of rubber involves several steps. First is the recipe selection with proper ingredients and correctly weighing each ingredient. The selected formulation should be easily processable, i.e., mixing, calendaring and extrusion with maximum scorch safety. It can be then molded and cured within a reasonable time period to produce a specific product with desired properties. These steps should be performed at the lowest possible cost. Thus, rubber compounding has three very important criteria, which the author calls three P's; they are processing, properties and price.

Major compounding ingredients and their functions:

(A) **Elastomers**—This is the most important ingredients in rubber formulation. Specific elastomers are selected for desired compound properties.

(B) **Fillers**—These are used to reinforce or enhance properties of elastomers while reducing cost of the compound. In black compounds, carbon blacks are used. For white compound, silica, clay, calcium carbonate, etc. can be used.

(C) **Processing Aids**—These materials are used to help in mixing, calendaring, extrusion and molding by lowering the viscosity of a compound. Examples are various oils and plasticizers.

橡胶配合	
硫化	
老化/弹性体	
配合剂	
硫化	
配方	
混炼/压延	
挤出	
标准	
补强	
胶料	
炭黑/白炭黑/黏土/碳酸钙	
加工助剂	
成型/黏度	
增塑剂	

(D) Antidegradants—These chemicals are used to protect rubber both in uncured and cured states, from oxidation, ozonization and aging and, therefore, aid in extending product life. 抗降解剂 氧化/臭氧化

(E) [2] Vulcanizing Agents—These chemicals, upon heating, crosslink elastomer molecules to provide harder, more thermally stable elastic products. Sulfur is the primary vulcanization agent. However, in some cases, peroxides are also used. Curing, vulcanization, and crosslinking are synonymous and are used interchangeably. 硫化剂 交联 硫黄 过氧化物

(F) Accelerators—These materials accelerate the vulcanization by increasing the rate of crosslinking reactions. 促进剂 交联反应

(G) Activators—These chemicals form complexes with accelerators and further active the curing process. Zinc oxide and stearic acid are commonly used activators. 氧化锌/硬脂酸

[3] In addition, special materials, not normally used in rubber compounds, may be needed to impart certain characteristics to compounds. Examples are: blowing agents, fabric-rubber adhesion promoters, tackifiers and flame retardants. 特殊材料 发泡剂/织物-橡胶黏合促进剂/增黏剂 阻燃剂

New Words

elastomer	[ɪˈlæstəmə(r)]	n.	弹性体
cure	[kjʊə(r)]	v.	硫化
curing	[ˈkjʊərɪŋ]	n.	硫化
vulcanize	[ˈvʌlkənaɪz]	v.	硫化
vulcanization	[ˌvʌlkənɪˈzeɪʃən]	n.	(橡胶的)硫化(过程)
crosslink	[ˈkrɒslɪŋk]	n.	交联
crosslinking	[kˈrɒslɪŋkɪŋ]	v.	交联
ingredient	[ɪnˈgriːdiənt]	n.	配合剂
recipe	[ˈresəpi]	n.	配方
formulation	[fɔːmjʊˈleɪʃn]	n.	配方
calendar	[ˈkælɪndə(r)]	n.	压延/压延机
extrusion	[ɪkˈstruːʒn]	n.	挤出
processing	[ˈprəʊsesɪŋ]	n.	加工
property	[ˈprɒpəti]	n.	性能
filler	[ˈfɪlə(r)]	n.	填料
plasticizer	[ˈplæstɪsaɪzə]	n.	增塑剂
sulfur	[ˈsʌlfə]	n.	硫黄
synonymous	[sɪˈnɒnɪməs]	adj.	同义的;同义词的
interchangeably	[ɪntəˈtʃeɪndʒəblɪ]	adv.	可交换地,可交替地

Notes

[1] Rubber compounding includes selection of proper ingredients to enable one to process

compound and vulcanize to give desirable physical and chemical properties, especially after aging. 橡胶配合包括选择合适的配合剂，使人们能够对胶料进行加工和硫化；赋予胶料（特别是老化后的胶料）理想的物理和化学性能。

[2] Vulcanizing Agents—These chemicals, upon heating, crosslink elastomer molecules to provide harder, more thermally stable elastic products. 硫化剂：加热后，硫化剂可以交联弹性体分子，以提供硬度更好、热稳定更高的弹性制品。

[3] In addition, special materials, not normally used in rubber compounds, may be needed to impart certain characteristics to compounds. Examples are: blowing agents, fabric-rubber adhesion promoters, tackifiers and flame retardants. 另外，某些通常不用于橡胶配合的特殊材料，可用来赋予胶料某些特性。例如：发泡剂，织物-橡胶黏合促进剂，增黏剂和阻燃剂等。

Exercises

1. Translate the first and second paragraphs of the text into Chinese.
2. Put the following words into English.

橡胶配合	配合剂	弹性体	硫化	硫化剂
混炼	压延	挤出	促进剂	活性剂
增塑剂	抗降解剂	增黏剂	发泡剂	阻燃剂

[Reading Material]

History and Causes of Rubber Compound

The rubber compound was first developed by Goodyear and Hancock and it continues to develop as new materials and new variations on old ones appear in the marketplace. The compound we see everyday as rubber, such as in a tire or pencil eraser, is a mixture of a number of different ingredients. It starts with the raw gum elastomer, supplied by the plantation owner as NR, or by the petrochemical complex converting petroleum products such as ethylene, propylene and butadiene into "raw" bales or chips of rubbery polymers such as EPDM, BR, SBR, NBR or CR. It is shipped to the rubber processor who blends it with various ingredients. The raw gum elastomer itself has very limited use, although adhesives provide one example. Most are mechanically weak and subject to significant swelling in liquids, and will not retain their shape after molding. Many of its other properties could also benefit from enhancement. It is at this point that the rubber compounder takes over, and all of his art and science is dedicated to modifying the raw gum elastomer, changing it into a more useful material.

Words and Expressions

marketplace	[ˈmɑːkɪtpleɪs]	n.	市场，业界
tire	[ˈtaɪə(r)]	n.	轮胎
gum	[gʌm]	n.	树胶；黏胶
raw gum elastomer			生胶弹性体
plantation owner			种植园主

petrochemical	[ˌpetrəʊˈkemɪkl]	n/adj.	石油化学品/石油化学的
petroleum	[pəˈtrəʊliəm]	n.	石油
rubbery	[ˈrʌbəri]	adj.	橡胶似的,有弹性的
swelling	[ˈswelɪŋ]	n.	膨胀
molding	[ˈməʊldɪŋ]	n.	成型
be dedicated to			致力于

Lesson 16 Compound Formulation

Recipe selection— [1] Compounding starts with selection of proper ingredients to enable one to mix, process it further and vulcanize to give desired properties. Ingredients in compound formulations (also called recipes) are based on per hundred part of rubber (phr) and are usually listed in the order of addition in mixing. A typical passenger tire tread recipe is listed in Table 3.1.

Table 3.1 Passenger tire tread recipe

item	phr	Approximate/lb①	Total price of ingredient
SBR 1500	60.0	1.38	$82.80
cis-Polybutadiene	40.0	1.62	$64.80
Carbon black	70.0	0.68	$47.60
Aromatic oil	37.5	0.39	$14.60
Zinc oxide	5.0	0.87	$4.35
Stearic acid	1.0	0.43	$0.43
TMQ (antioxidant)	2.0	2.48	$4.96
6PPD (antiozonant)	1.5	2.66	$3.99
Sulfur	1.5	0.18	$0.27
Sulfenamide accelerator	1.0	3.21	$3.21
TMTD	0.2	2.87	$0.84
Total weight/price	219.7②		$227.84

① Retail prices as of 9/27/99.
② Price per lb=227.84/219.7=$1.037.

Calculations of mixing batch sizes—To calculate batch size of a typical laboratory B-Banbury, consider the total weight of the above recipe as 219.7 grams. Banbury volume is 1500cm³, and fill factor is 0.80. Hence, total weight of batch for B-Banbury will be $= 1500 \times 1.12 \times 0.80 = 1344$g. The specific density of compound is 1.12g/cm³.

Cost calculations—Calculate price of each ingredient for the actual amount used. Then get total weight of all ingredients (say in pounds) and total cost of all ingredients. By dividing total cost by total weight (in pounds), one can get the compound cost/lb. Cost per lb = Total cost of all ingredients/Total recipe weight in pounds That cost is only for the materials used and does not include any other cost, e.g., mixing, processing.

Moreover, of the ingredients used in large quantities in a compound, elastomers are probably the most expensive materials. [2] Carbon black and oils, that help improve uncured and cured compound's properties, also help lower compound cost. Therefore, a proper balance between compound properties and compound cost should be achieved for a rubber product.

New Words

batch	[bætʃ]	n.	一批
density	['densəti]	n.	密度
aromatic	[ˌærə'mætɪk]	adj.	芳香族的
stearic	[stɪ'ærɪk]	adj.	硬脂酸的
stearic acid			硬脂酸
zinc	[zɪŋk]	n.	锌
oxide	['ɒksaɪd]	n.	氧化物
zinc oxide			氧化锌
pound	[paʊnd]	n.	磅

Notes

[1] Compounding starts with selection of proper ingredients to enable one to mix, process it further and vulcanize to give desired properties. 胶料配合首先需选择适当的配合剂以使其能够混炼、加工、硫化进而获得期望的性能。

[2] Carbon black and oils, that help improve uncured and cured compound's properties, also help lower compound cost. 炭黑和油（增塑剂）有助于提高未硫化胶和硫化胶性能，也有助于降低胶料成本。

Exercises

1. Translate the first and second paragraphs of the text into Chinese.
2. Put the following words into English.

配合剂　　　　配方选择　　　　胎面　　　　氧化锌　　　　硬脂酸
填充因子　　　每百份橡胶　　　胶料性能　　胶料成本　　　成本计算

[Reading Material]

The Basic Compound Formula

The specific formulation in Table 3.2 has 100 parts of raw gum elastomer and 160.15 parts of total material. After curing for 35 minutes at 140℃, its vulcanized properties are indicated as 57 IRHD, with a tensile strength of 30MPa and elongation at break of 645%.

Rubber chemists use the term phr (parts per hundred rubber), meaning parts of any nonrubbery material per hundred parts of raw gum elastomer (rubbery material). They prefer this rather than expressing an ingredient as a percentage of the total compound weight. Parts can mean any unit of weight (kg, lb, etc.) as long as the same weight unit is used throughout the formulation. The compound formulation in Table 3.2 is a typical one in the rubber industry for most unsaturated (sulfur cross-linked) elastomers.

The raw gum elastomer is the key ingredient (the one which is actually cross-linked) on which depend many of the properties of the final product. It is therefore always at the top of the formulation list and is expressed as 100 parts by weight of the total recipe. Thus, in the formulation in Table 3.2, for a 100kg of raw gum elastomer there will always be 5kg of zinc oxide.

Table 3.2 A Basic compound formula

Material	phr	Specific formulation for example	phr
Raw gum elastomer	100	SMR 20	100
Sulfur	0~4	Sulfur	0.35
Zinc oxide	5	Zinc oxide	5
Steric acid	2	Stearic acid	2
Accelerators	0.5~3	MBS	1.4
		TMTD	0.4
Antioxidant	1~3	HPPD	2
Filler	0~150	N330 Black	45
Plasticizer	0~150	Aromatic petroleum oil	4
Miscellaneous		None	
		TOTAL	160.15

Words and Expressions

raw gum elastomer			生胶
rubbery	['rʌbəri]	adj.	橡胶似的,有弹性的
miscellaneous	[ˌmɪsə'leɪniəs]	adj.	各种各样的;其他配合剂

Lesson 17　Raw gum elastomer

Monomers are polymerized to produce macromolecules or polymers.　　单体/聚合/大分子

[1] Elastomers are those polymers, usually with double bonds in the backbone, that can be vulcanized to give elastic properties. All elastomers are polymers but not all polymers are elastomers.

Elastomers can be grouped into three categories:

(1) General purpose rubbers;　　通用橡胶
(2) Solvent resistant rubbers;　　耐溶剂橡胶
(3) Heat resistant rubbers.　　耐热橡胶

Table 3.3　Three different group of elastomers

Type	Rubber	Monomers used	Application
General purpose	Nature rubber	Isoprene	Tires, belts, hoses
	Synthetic polyisoprene	Isoprene	
	Polybutadiene	Butadiene	
	Styrene-butadiene	Styrene, butadiene	
	Butyl	Isobutylene, isoprene	
	Ethylene-propylene-diene Rubber (EPDM)	Ethylene, propylene and diene monomer	
Solvent resistant	Nitrile	Butadiene, acrylonitrile	Oil seals, oil hoses automotive gaskets
	Neoprene	Chloroprene	
Heat resistant	Silicone	Dimethylsiloxane	Spark plug cables gaskets, seals
	Polyacrylate	Acrylate	
	Fluorocarbon	Tetrafluoroethylene	

油封/输油软管
汽车垫圈
火花塞电缆
垫圈/密封制品
四氟乙烯

Examples of these are included in Table 3.3. [2] If one elastomer cannot provide desired set of properties, several different elastomers with proper cure compatibility are blended together. An example will be the use of natural rubber (NR) for a product that needs good tensile properties, tear strength and ozone resistance. NR basically gives high tensile strength, excellent tear strength but poor ozone resistance. Thus, to improve ozone resistance, it can be blended with EPDM which has excellent ozone resistance. Detail introductions of each raw rubber are discussed in Part B in this book.

硫化相容性/混合
拉伸性能
撕裂强度/耐臭氧性

New Words

monomer	['mɒnəmə]	n.	单体
polymer	['pɒlɪmə(r)]	n.	[高分子]聚合物
polymerize	['pɒlɪməraɪz]	v.	(使)聚合
polymerization	[ˌpɒlɪmaɪ'zeɪʃn]	n.	聚合
compatibility	[kəmˌpætə'bɪləti]	n.	相容性/相通性
isoprene	['aɪsəpriːn]	n.	异戊二烯
butadiene	[ˌbjuːtə'daɪiːn]	n.	丁二烯

isobutylene	[aɪsəʊˈbjuːtɪliːn]	n.	异丁烯
styrene	[ˈstaɪriːn]	n.	苯乙烯
ethylene	[ˈeθɪliːn]	n.	乙烯
propylene	[ˈprəʊpəliːn]	n.	丙烯
acrylonitrile	[ˌækrələʊˈnaɪtrɪl]	n.	丙烯腈
chloroprene	[ˈklɔːrəpriːn]	n.	氯丁二烯
acrylate	[ˈækrɪleɪt]	n.	丙烯酸酯
tire	[ˈtaɪə(r)]	n.	轮胎
belt	[belt]	n.	传送带
hose	[həʊz]	n.	软管,胶皮管

Notes

[1] Elastomers are those polymers, usually with double bonds in the backbone, that can be vulcanized to give elastic properties. All elastomers are polymers but not all polymers are elastomers. 弹性体是能够被硫化并赋予弹性性质的聚合物,这些聚合物骨架中通常含有双键。所有弹性体都是聚合物,但并非所有聚合物都是弹性体。

[2] If one elastomer cannot provide desired set of properties, several different elastomers with proper cure compatibility are blended together. 如果一种弹性体不能提供所需的一组性能,则可将几种具有适当硫化相容性的不同弹性体混合在一起。

Exercises

1. Translate the last paragraph of the text into Chinese.
2. Put the following words into English.

单体	聚合物	通用橡胶	耐溶剂橡胶	耐热橡胶
油封	输油软管	储罐衬里	火花塞电缆	垫圈
密封制品	硫化相容性	拉伸性能	撕裂强度	耐臭氧性

[Reading Material]

Compounding Properties and Application of Several Rubbers

(1) NR, SBR, BR—Compounding, processing and curing of NR, SBR and polybutadiene (BR) are similar. Polybutadiene is generally used in blends with NR and SBR to improve abrasion resistance of the blend. For compounds containing these elastomers, normal ingredients used are: fillers, processing aids, plasticizers, antioxidants, antiozonants, vulcanizing agents, acceleration and activators. SBR and BR require less sulfur than NR and more accelerators for same state of cure. Emulsion SBR is slower curing than solution SBR. Broader molecular weight distribution (MWD), as in NR and emulsion SBR, is good for processability. Solution SBR and BR have, in general, narrower MWD, and, hence, are more difficult to process.

Unlike NR, most synthetic elastomers require tackifiers to impart good building tack. Major uses of NR are in tires, engine mounts and other mechanical goods. SBR and BR are primarily used in tires. A small percentage of SBR is used in mechanical goods. Some BR is used in golf balls and for impact modification of plastics.

(2) Ethylene propylene diene monomer (EPDM)—These elastomers have un-saturation dangling from the main backbone and, hence, are more resistance to ozone attack. The unsaturation levels are usually in the 2% ~ 10% range. EPDM with high unsaturation are more cure compatible with diene rubbers (e.g. NR, SBR, BR) and are generally used to improve ozone resistance. EPDM can generally tolerate higher loadings of filler and oil. EPDM can be cured using sulfur-accelerator systems. Such elastomers are used in roofing materials, automotive insulation and weather-stripping, tire sidewalls, and wire and cables.

(3) Nitrile rubber—This rubber is made of two co-monomers: acrylonitrile (ACN) and butadiene (BD). Relative contents of these monomers determine elastomer properties. The higher the BD content, the lower the T_g (glass transition temperature) and better the low temperature properties. On the other hand, higher ACN content is needed for better oil resistance. Significant improvement in heat resistance is obtained by hydrogenation of nitrile rubber thus eliminating most of the double bonds.

Nitrile rubber can be compounded like SBR. Conventional major ingredients are needed. Main uses are in oil seals, gaskets, printing rolls, adhesives, shoe products and coated fabrics.

(4) Poly-chloroprene—Most of poly-chloroprenes have *trans*-1, 4 configuration and crystallize easily to give high tensile strength. The chlorine content provides the oil resistance feature.

Vulcanization of neoprene is done primarily with use of metal oxide such as zinc oxide and magnesium oxide. These two are sufficient to cure without much use of convention accelerators. Processing aids such as peptizers or oils are used with G-type neoprene. Carbon blacks, mainly thermal and furnace blacks, and mineral fillers are used. Silica provides higher tensile strength and hot tear resistance. Because poly-chloroprenes have residual double bond in the backbone, at least 2 phr antioxidant is needed to improve heat resistance.

The main uses of chloroprene are in footwear, wire and cable, V-belts and hoses.

(5) Silicone rubber—Silicone rubbers have excellent heat resistance, and have good resistance to oxygen and ozone because of saturated backbone. They also have good low temperature flexibility, and good compression set. Basic silicone compound can be very simple: silicone rubber, silica filler and organic peroxide as a curing agent. Very fine particle silica gives improved tensile and tear strength. Semi-reinforcing white fillers such as calcium carbonate is used to lower cost with moderate tensile strength.

Major uses of silicone rubber are in automotive ignition cables, coolant and heater hoses, gaskets, adhesives. O-rings, seals and gaskets made of silicone rubber are used in aerospace industry and in appliances.

Words and Expressions

compounding	[ˈkɒmpaʊndɪŋ]	n.	配合
abrasion	[əˈbreɪʒn]	n.	磨损
abrasion resistance			耐磨性
molecular weight distribution			分子量分布

tackifier	['tækɪfaɪə]	n	增黏剂
building tack			黏性
engine mount			发动机支架
golf ball			高尔夫球
modification	[ˌmɒdɪfɪ'keɪʃn]	n.	改进,改性
roofing material			屋顶材料
automotive insulation			汽车绝缘材料
tire sidewall			胎侧
acrylonitrile	[ˌækrələʊ'naɪtrɪl]	n.	丙烯腈
glass transition temperature			玻璃化转变温度
oil resistance			耐油性
heat resistance			耐热性
printing roll			印刷胶辊
coated fabric			涂层织物
oil seal			油封
gasket			垫片
configuration	[kənˌfɪgə'reɪʃn]	n.	构象
crystallize	['krɪstəlaɪz]	v.	(使)结晶
metal oxide			金属氧化物
peptizer	['peptaɪzər]	n.	塑解剂
hot tear resistance			抗热撕裂性
oxygen	['ɒksɪdʒən]	n.	氧,氧气
ozone	['əʊzəʊn]	n.	臭氧
organic peroxide			有机过氧化物
automotive ignition cable			汽车引擎电缆
coolant hose			冷却软管
O-ring			O形圈
aerospace industry			航空工业

Lesson 18 Vulcanizing system

Vulcanizing system include sulfur, accelerators and activators for sulfur-based cure systems. The most common vulcanizing agent is sulfur. During curing, a three-dimensional cross-linked network is formed which imparts properties to compounds. [1] The extent and rate of vulcanization is measured by using a cure-meter which gives a curve plotting modulus as a function of cure time. Figure 3.1 is a typical rheometer curve. This type of curve gives information on minimum torque, maximum torque, scorch time (T_s) and t_c (90) (cure time to reach 90% of maximum torque).

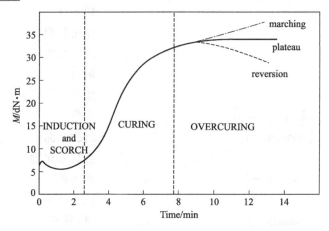

Figure 3.1 Typical rheometer curve

[2] In general, crosslink density, which is a measure of the extent of vulcanization, increases with cure time. Compound properties are functions of crosslink density. Figure 3.2 shows effects of crosslink density on tensile strength, tear strength, fatigue life, hardness and hysteresis. Crosslink density and type of crosslinks both affect compound properties.

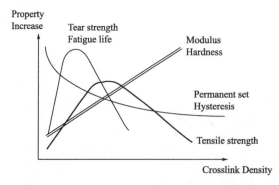

Figure 3.2 Effect of crosslink density on properties

The higher the sulfur content, the more polysulfidic the crosslinks.

Polysulfidic crosslink gives poor aging properties. Mono-sulfidic or di-sulfidic crosslinks (efficient vulcanization, EV) on the other hand, provide poor fatigue life. Therefore, compounds are designed to give the best overall properties with minimum trade-offs.

[3] Sulfur alone takes a commercially prohibitive length of time to cure a rubber compound. Therefore, chemical accelerators are used to speed up the cure rate. Additionally, activators are used to further active the curing process. Table 3.4 is an example of natural rubber compound showing effects of an activator and accelerator. An activator reduced the cure time, but still it is too long. The accelerator made it much faster.

Table 3.4 Effect of activator and accelerator on cure times of NR compounds

No.	NR compounds	Cure time at 140℃
1	100 phr NR 8 phr sulfur	8 hours
2	100 phr NR 8 phr sulfur 3 phr zinc oxide	3 hours
3	100 phr NR 8 phr sulfur 3 phr zinc oxide 1.5 phr CBTS(Accelerator)	12 minutes

Table 3.5 Chemical classification of accelerators

Type	Example	Typical use
Aldehyde-amine reaction product	Butyraldehyde-aniline condensation product	Self-curing adhesives
Amines	Hexamethylene tetramine	Delayed action for NR
Guanidines	Diphenyl guanidine	Secondary accelerator
Thioureas	Ethylenethiourea	Fast curing for CR
Thiazoles	2-Mercaptobenzothiazole	Fast curing, general-purpose, broad curing range
	Benzothiazole disulfide	Safe processing, general-purpose, moderate cure rate
Thiurams	Tetramethylthiuram disulfide	Safe, fast curing
Sulfenamides	N-cyclohexyl-2-benzothiazylsulfenamide	Safe processing, delayed action
Dithiocarbamates	Zinc dimethyldithiocarbamate	Fast, low temperature use
Xanthates	Dibutylxanthogen disulfide	General-purpose, low temperature use
	Zinc isopropyl xanthate	Latex and adhesives room-temperature curing

General classification of accelerators are listed in Table 3.5.

[4] Although the use of an accelerator is about 1~2 phr in a compound, it has a major effect on cure rate and properties of a compound. Zinc oxide and stearic acid are most commonly used activators. Activators form complexes with accelerators making them more effective in

increasing cure rate. Zinc oxide usage is about 3～5 phr, whereas stearic acid is used at 1～2 phr.

[5] Elastomers with no or low unsaturation are generally cured with <u>peroxides</u>. Such crosslinks are C—C type and are thermally stable. Peroxides cure, therefore, give improved <u>compression set</u>, lower <u>creep</u> and <u>stress relaxation</u>. Peroxide cures can be used in high <u>transparency</u> or non-discoloring compounds. Poor <u>flex life</u> and low tear strength are major disadvantages of peroxide cures.

过氧化物
压缩永久变形/蠕变
应力松弛/透明度
屈挠寿命

New Words

sulfur	['sʌfə]	n.	硫黄
accelerator	[ək'seləreɪtə(r)]	n.	促进剂
activator	['æktɪveɪtə]	n.	活化剂
cure	[kjʊə(r)]	v.	硫化
curing	['kjʊərɪŋ]	n.	硫化
vulcanization	[ˌvʌlkənaɪ'zeɪʃn]	n.	硫化
modulus	['mɒdjʊləs]	n.	模量
rheometer	[rɪ'ɒmɪtə]	n.	流变仪
torque	[tɔːk]	n.	扭矩
scorch	[skɔːtʃ]	n.	焦烧
effective	[ɪ'fektɪv]	adj.	有效的
classification	[ˌklæsɪfɪ'keɪʃn]	n.	分类;分级
aldehyde-amine		n.	醛胺
amine	[ə'miːn]	n.	胺
guanidine	['gwænədiːn]	n.	胍
thiourea	[θaɪəʊjʊə'rɪə]	n.	硫脲
thiazole	['θaɪəˌzəʊl]	n.	噻唑
thiuram	['θɪjʊræm]	n.	秋兰姆
sulfenamide	[sʌl'fenəmaɪd]	n.	次磺酰胺
dithiocarbamate	[daɪˌθaɪəʊ'kɑːbəmeɪt]	n.	二硫代氨基甲酸盐
xanthate	['zænθeɪt]	n.	黄原酸盐
peroxide	[pə'rɒksaɪd]	n.	过氧化物
creep	[kriːp]	n.	蠕变
stress	[stres]	n.	应力
strain	[streɪn]	n.	应变
relaxation	[ˌriːlæk'seɪʃn]	n.	松弛
transparency	[træns'pærənsi]	n.	透明度

Notes

[1] The extent and rate of vulcanization is measured by using a cure-meter which gives a

curve plotting modulus as a function of cure time. 硫化的程度和速率可用硫化仪来测量，该硫化仪给出胶料模量与硫化时间关系的曲线图。

［2］In general, crosslink density, which is a measure of the extent of vulcanization, increases with cure time. Compound properties are functions of crosslink density. 通常，交联密度（表征硫化程度）随硫化时间的延长而增加。胶料性能是交联密度的函数。

［3］Sulfur alone takes a commercially prohibitive length of time to cure a rubber compound. Therefore, chemical accelerators are used to speed up the cure rate. Additionally, activators are used to further active the curing process. 单独使用硫黄硫化胶料需要很长的时间。因此，可采用化学促进剂加快硫化速率。此外，活化剂用于进一步活化硫化工艺过程。

［4］Although the use of an accelerator is about 1～2phr in a compound, it has a major effect on cure rate and properties of a compound. 尽管在胶料中，促进剂的用量约为1～2phr，但它对胶料的硫化速率和硫化胶性能有重要影响。

［5］Elastomers with no or low unsaturation are generally cured with peroxides. Such crosslinks are C—C type and are thermally stable. Peroxides cure, therefore, give improved compression set, lower creep and stress relaxation. 饱和或不饱和度低的弹性体通常用过氧化物硫化。过氧化物硫化生成的交联键是C—C键且热稳定性较高。因此，过氧化物硫化可以改善胶料的压缩永久变形，降低蠕变和应力松弛。

Exercises

1. Translate the first and fourth paragraphs of the text into Chinese.
2. Put the following words into English.

硫黄	硫化剂	促进剂	活化剂	防焦剂
氧化锌	硬脂酸	硫化体系	硫黄硫化体系	三维交联网络
硫化仪	最大扭矩	最小扭矩	焦烧时间	工艺正硫化时间
硫化时间	交联密度	双硫交联键	多硫交联键	拉伸强度
撕裂强度	疲劳寿命	滞后损失	压缩永久变形	屈挠寿命

3. What is the component of vulcanizing system?

[Reading Material]

Introduction of Sulfur

It is interesting to note that sulfur is still by far the most used crosslinking agent in the rubber industry since its use by Goodyear and Hancock. It reacts chemically with the raw gum elastomer forming crosslinks between the polymer chains, resulting in a more dimensionally stable and less heat-sensitive product. Its cost is relatively low but its function is essential. It is available in different particle sizes (fineness), and can also have a small quantity of oil added to reduce its dust in the air during handling. Sulfur is suitable for vulcanizing rubber; it has a low ash content, low acidity and sufficient fineness for adequate dispersion and reaction. The finer particle sizes, coated with magnesium carbonate, assist its dispersion in elastomers such as nitrile. Sometimes, as the sulfur level in a compound is increased, some of it can slowly bloom to the surface. For example, Heinisch mentions that sulfur levels as low as around 1 phr (at room temperature) might bloom.

Blooming occurs if an additive dissolves totally in the polymer at the processing temperature but is only partially soluble at ambient temperature. In this situation, some of the additive precipitates out of solution on cooling collecting on the surface of the polymer mass, causing a bloom. In this case, a highly "polymeric" (amorphous) form of sulfur, known as insoluble sulfur, is available to reduce this problem, although dispersion in the compound can be more difficult. Although bloom does not generally affect a product's performance it is aesthetically displeasing. In the uncured compound bloom can reduce tack needed in building operations (such as plying up uncured sheets of rubber to obtain thicker sheets).

Words and Expressions

dimensionally stable			尺寸稳定的
heat-sensitive product			热敏制品
ash	[æʃ]	n.	灰,灰烬
coated	['kəʊtɪd]	adj.	涂上一层的
coated with			(被)包覆
magnesium	[mæg'niːziəm]	n.	[化]镁(金属元素)
magnesium carbonate			碳酸镁
bloom	[bluːm]	v./n.	大量出现/喷霜
blooming	['bluːmɪŋ]	n.	喷霜;喷硫
ambient	['æmbiənt]	adj.	周围的,环境
ambient temperature			环境温度
precipitate	[prɪ'sɪpɪteɪt]	v./n.	使沉淀/沉淀物
polymer mass			聚合物基体
amorphous	[ə'mɔːfəs]	adj.	无固定形状的,非结晶的
insoluble	[ɪn'sɒljəbl]	adj.	不溶的
insoluble sulfur			不溶性硫黄
aesthetically	[es'θetɪklɪ]	adv.	审美地,美学观点上地
displeasing	[dɪs'pliːzɪŋ]	adj.	不愉快的,令人发火的
uncured	[ʌn'kjuːəd]	adj.	未硫化的
tack	[tæk]	n.	黏性

Lesson 19 Fillers

Two types of fillers are commonly used by the rubber industry:

(1) Carbon black   Mainly used to reinforce rubber compounds, and to some extent, to lower compound cost.

(2) White fillers   Used in non-black compounds to semi-reinforce and to lower cost. Silica filler is used to provide reinforcement.

Carbon Blacks—Carbon blacks are produced by either furnace or thermal processes and are respectively known as furnace or thermal blacks. [1] In general, the smaller the particle size, the more reinforcing the black, i.e., improvement in tensile, modulus, hardness and abrasion resistance. Different ASTM blacks and their average particle sizes are given in Table 3.6. N-100 to N-300 are considered hard blacks, the rest are considered soft blacks.

Table 3.6 Classification of carbon blacks

ASTM number	Average particle size/nm	Old classification system
N100~199	11~19	SAF
N200~299	20~25	ISAF
N300~399	26~30	HAF, EPC
N500~599	40~48	FEF
N600~699	49~60	GPF
N700~799	61~100	SRF
N900~999	201~500	MT

[2] Mooney viscosity of uncured compounds, moduli of cured compounds are depends on carbon black structure and loading. The higher the loading and the finer the particle size, the higher the Mooney viscosity for uncured compounds.

Table 3.7 Effects of particle size and structure on compound properties

Compound properties	Decreasing particle size (constant structure)	Increasing structure (constant particle size)
Black incorporation time	Not important	—
Dispersion time	—	+
Extrusion shrinkage and die swell	Not important	+
Viscosity	+	+
Green strength	+	+
Hardness	+	+
Tensile strength	+	Variable
Modulus	+	+
Elongation	—	—
Resilience	—	Not important

It should be pointed out, however, that the finer particle carbon blacks are harder to disperse and require more energy for mixing. They also increase hysteresis of compounds. Table 3.7 summarized effects of particle size and structure of carbon black on select properties.

Non-black Fillers—Common non-black fillers, their average particle sizes and functions are listed in Table 3.8.

[3] Non-black fillers contribution to compound properties depends on the surface area. High surface area filler provides higher reinforcement but are harder to disperse. For silica, higher surface area results in longer cure time and higher heat build-up. It appears that about 150m^2/g is optimum particle size for abrasion resistance and tear strength. [4] For proper reinforcement with silica, silane coupling agent is added in the first stage of mixing cycle. This improves abrasion resistance and reduces heat build-up.

Major uses of silica are in OTR tires, some in passenger and truck tires. Some shoe sole compounds also use silica.

Clays are used in footwear, mechanical goods, tire innerliners and white sidewalls. Clays when modified with silanes show improved reinforcement. Clays are grouped into "hard" and "soft" classification. In general, hard clays have smaller particle sizes and provide semi-reinforcement. Most clays are acidic and slow the cure rates of sulfur-accelerator cure systems. Calcium carbonate, on the other hand, is basic and accelerates the cure rate.

Table 3.8 Average particle sizes of non-black filler

Fillers	Average particle size/μm	Function
Clays	0.20~10	Semi-reinforcing
Precipitated calcium carbonate	0.05~1	Extender
Natural calcium carbonate	1~20	Extender
Precipitated silica	0.02~0.10	Reinforcing

New Words

filler	['fɪlə(r)]	n.	填料
reinforce	[ˌriːɪn'fɔːs]	v.	加强,补强
reinforcement	[ˌriːɪn'fɔːsmənt]	n.	加强,补强
furnace	['fɜːnɪs]	n.	熔炉,火炉
silica	['sɪlɪkə]	n.	二氧化硅,白炭黑
thermal	['θɜːml]	adj.	热的
hardness	['hɑːdnəs]	n.	硬度
abrasion	[ə'breɪʒn]	n.	磨损
resistance	[rɪ'zɪstəns]	n.	抵抗;抗,耐
viscosity	[vɪ'skɒsəti]	n.	黏性;黏度
moduli	['mɒdʒəˌlaɪ]	n.	模量(modulus 复数)

modulus	['mɒdjʊləs]	n.	模量
shrinkage	['ʃrɪŋkɪdʒ]	n.	收缩
swell	[swel]	n.	膨胀
elongation	[ˌiːlɒŋ'geɪʃn]	n.	伸长,伸长率
resilience	[rɪ'zɪliəns]	n.	回弹性
hysteresis	[ˌhɪstə'riːsɪs]	n.	滞后损失
clay	[k'leɪ]	n.	陶土
silane	['sɪleɪn]	n.	硅烷
footwear	['fʊtweə(r)]	n.	(总称)鞋类
acidic	[ə'sɪdɪk]	adj.	酸的,酸性的
basic	['beɪsɪk]	adj.	碱的,碱性的
accelerate	[ək'seləreɪt]	v.	加快,增速

Notes

[1] In general, the smaller the particle size, the more reinforcing the black, i. e., improvement in tensile, modulus, hardness and abrasion resistance. 通常，炭黑的粒径越小，补强效果越好，即拉伸强度、模量、硬度和耐磨性的改善。

[2] Mooney viscosity of uncured compounds, moduli of cured compounds are depends on carbon black structure and loading. The higher the loading and the finer the particle size, the higher the Mooney viscosity for uncured compounds. 未硫化胶料的门尼黏度以及硫化胶料的模量取决于炭黑的结构和填充量。填充（炭黑）含量越高，（炭黑）粒径越细，未硫化胶料的门尼黏度越高。

[3] Non-black fillers contribution to compound properties depends on the surface area. High surface area filler provides higher reinforcement but are harder to disperse. For silica, higher surface area results in longer cure time and higher heat build-up. 非黑（浅色）填料对胶料性能的影响取决于填料比表面积。高比表面积填料赋予（胶料）更高的强度，但较难分散。对于白炭黑，其较高的比表面积导致硫化时间延长和热量积聚增加。

[4] For proper reinforcement with silica, silane coupling agent is added in the first stage of mixing cycle. This improves abrasion resistance and reduces heat build-up. 为得到合适白炭黑补强效果，在混炼的第一阶段加入硅烷偶联剂，这样能提高胶料耐磨性并减少了热量积聚。

Exercises

1. Translate the fourth and sixth paragraphs of the text into Chinese.
2. Put the following words into English.

填料	炭黑	白炭黑	陶土	沉淀法碳酸钙
补强	半补强	门尼黏度	硬度	耐磨性
粒子尺寸	比表面积	热量集聚	模量	硅烷偶联剂
工程机械轮胎	鞋底胶料	轮胎内衬	回弹性	分散

3. What is the effect of particle size of carbon black on reinforcing effect of rubber?

[Reading Material]

Introduction of Carbon Black and Precipitated Silica

Carbon black

  This is a material of major significance to the rubber industry, so it is no surprise that most rubber products we see in the market place are black in color. We have moved a long way from collecting carbon from smokey oil flames, which produced a material called lampblack. The next historical step was to burn natural gas against iron channels, then scrape off the carbon to produce a highly reinforcing material called channel black. Both the use of this black in the rubber industry and its source of supply are currently limited and its cost is somewhat high. There are two common methods of producing carbon black today. Heating natural gas in a silica brick furnace to form hydrogen and carbon, produces a moderately reinforcing material called thermal black. Alternatively, if we incompletely burn heavy petroleum fractions, then furnace blacks are produced. These are the most important blacks in terms of quantity used and available types.

  Carbon black consists of extremely small particles (from around 10 to 300nm) in a grapelike aggregate. This gives two primary properties allowing a whole range of grades designated by both a particular particle size (surface area) and a specific level of structure. The rubber compounder thus has a whole range of properties available to him. The American Society for Testing and Materials (ASTM) specifies generic codes for these grades (ASTM D 1765). Numbers after the letter N in this ASTM code relate to particle size but they do not relate to structure. For example, one such code name is an oil furnace black type called N110, which is a grade with a very small particle size (therefore highly reinforcing) and fairly high structure. An example of a thermal black type is N990. It has a large particle size, and low structure resulting in a much lower level of reinforcement but higher resilience.

  A decrease in carbon black particle size increases the tensile strength of the cured vulcanizate. For example, Cabot Corporation literature, illustrates an SBR elastomer with 20 phr of oil. Its compound is adjusted to equal vulcanized hardness of 65 Shore A by adjusting the loading level of each black. It gives a tensile strength at break of around 17.9 MPa for an N550 black and about 22.8 MPa for an N347 black. Both blacks have approximately the same structure but the N347 has the smaller particle size, and therefore gives the higher tensile strength at break. Carbon black is also a powerful UV absorber and therefore will give a measure of protection against sunlight to the rubber. This is especially important for unsaturated elastomers such as NR and SBR.

Precipitated silica

  Silica is a material found in abundance, to the delight of small children (and some adults) building sandcastles by the sea. The humble shoe sole probably marked the beginning of the use of silica in the rubber industry. Like carbon black, it reinforces raw gum elastomers, and by virtue of its white color, it does not impose a restriction on the color of the vulcanized products. Since silica is white, any color can be mixed in and be seen in the compound.

  For some properties it has the edge over carbon black. For example, it improves the tear strength of the vulcanized product and also better heat aging is claimed. However, it does not offer carbon black's wide range of grades. Precipitated silica can be associated with certain unusual processing and curing characteristics within the rubber compound. Stiff and boardy uncured compounds may result from higher filler levels. Also, its addition to a rubber compound requires greater accelerator levels for adequate cure, although this situation is somewhat mitigated by addition of triethanolamine (TEA) or diethylene glycol. Recent introduction of chemicals such as organosilanes, added to the compound, produce a lower mixed viscosity and an improvement in mechanical and some dynamic properties. Although silica is more expensive than carbon black, there is a huge supply of the raw material in nature.

Words and Expressions

flame	[fleɪm]	n.	火焰
lampblack	[ˈlæmpˌblæk]	n.	油炉法炭黑
iron channel			槽铁
brick	[brɪk]	adj.	用砖做的,似砖的
hydrogen	[ˈhaɪdrədʒən]	n.	氢
thermal black			热裂解炭黑
petroleum	[pəˈtrəʊliəm]	n.	石油
furnace	[ˈfɜːnɪs]	n.	熔炉,火炉
furnace black			炉法炭黑
grapelike	[gˈreɪplaɪk]	adj.	葡萄样的
aggregate	[ˈægrɪgət]	n.	团聚体;聚集体
particular	[pəˈtɪkjələ(r)]	adj.	特别的,独有的
particle size			粒度,粒径
American Society for Testing and Materials (ASTM)			美国材料试验协会
Cabot Corporation			卡博特公司
absorber	[əbˌsɔːbə]	n.	吸收体
abundance	[əˈbʌndəns]	n.	丰富,充裕
sandcastle	[ˈsændkɑːsl]	n.	沙堡
restriction	[rɪˈstrɪkʃn]	n.	限制,限定
edge	[edʒ]	n.	优势
stiff	[stɪf]	adj.	坚硬的
mitigate	[ˈmɪtɪgeɪt]	v.	减轻,缓和
triethanolamine	[traɪeθəˈnɒləmiːn]	n.	三乙醇胺
diethylene	[dɪəˈθɪliːn]	n.	二亚乙基
diethylene glycol			二甘醇
organosilane	[ɔːgənəʊˈzaɪlən]		有机硅烷

Lesson 20　Plasticizers

　　Oils and other "slippery" materials are called plasticizers (a somewhat vague term). ASTM D1566 defines them as "a compounding material used to enhance the deformability of a polymeric material". Their function at low levels is to aid in processing, e.g., mixing, calendaring, extrusion and molding by reducing the viscosity. [1] At higher amounts they reduce uncured compound viscosity, often lower compound cost, reduce vulcanite stiffness (hardness) and in some cases improve low temperature flexibility. [2] They also improve flow in extrusion and molding by making the uncured compound less elastic and reducing viscosity and friction.

　　Petroleum oils are one of the major sources of plasticizers. These oils are divided into three chemical categories, aromatic, naphthenic and paraffinic. [3] Some researchers point out that the latter category gives better rebound resilience and lower hysteresis, while aromatics are better for tensile strength and resistance to crack growth. It comes as no surprise that those elastomers which have little or no oil (petroleum) resistance are the ones most suited for compounding with petroleum oils.

　　For oil resistant elastomers such as NBR, liquid plasticizers such as esters (polar liquids) are used. Esters can also improve low temperature flexibility. [4] A few elastomers can hold large amounts of plasticizer and filler without appreciable degradation of properties, e.g., EPDM. Since a compound with large amounts of plasticizer can be difficult to mix, the compounder may purchase some raw gum elastomers with the plasticizer already mixed in (for example oil extended rubber). Chlorinated oils are used in some compounds to enhance flame retardation properties.

增塑剂
模糊的
可塑性
混炼/压延
挤出/成型
降低胶料成本

芳烃油类/环烷油类

回弹性
抗裂纹增长性

耐油弹性体

含氯增塑剂
阻燃性能

New Words

plasticizer	['plæstɪsaɪzə]	n.	增塑剂
vague	[veɪg]	adj.	模糊的
term	[tɜːm]	n.	术语
deformability	[dɪfɔːməˈbɪlɪtɪ]	n.	可塑性
flexibility	[ˌfleksəˈbɪlətɪ]	n.	柔韧性,可塑度
flow	[fləʊ]	n.	流动性
friction	[ˈfrɪkʃn]	n.	摩擦,摩擦力
aromatic	[ˌærəˈmætɪk]	adj.	芳香族的
naphthenic	[næfˈθiːnɪk]	adj.	环烷的
paraffinic	[pærəˈfɪnɪk]	adj.	石蜡族的
rebound	[rɪˈbaʊnd]	n.	回弹
ester	[ˈestə(r)]	n.	酯

appreciable	[əˈpriːʃəbl]	adj.	可估计的；相当可观的
degradation	[ˌdegrəˈdeɪʃn]	n.	恶化，降低

Notes

[1] At higher amounts they reduce uncured compound viscosity, often lower compound cost, reduce vulcanite stiffness (hardness) and in some cases improve low temperature flexibility. 增塑剂在用量较高情况下能降低未硫化胶料的黏度，降低胶料成本和硫化胶硬度，并在某些情况下能提高胶料的低温柔韧性。

[2] They also improve flow in extrusion and molding by making the uncured compound less elastic and reducing viscosity and friction. 增塑剂也能通过降低未硫化胶弹性、黏度和摩擦力进而改善胶料在挤出和成型加工过程中的流动性。

[3] Some researchers point out that the latter category gives better rebound resilience and lower hysteresis, while aromatics are better for tensile strength and resistance to crack growth. 一些学者指出后一类（石蜡油类）增塑剂能赋予胶料更好的回弹性和更低的滞后性，而芳香烃类增塑剂则（对胶料的）拉伸强度和抗裂纹增长有一定的改善。

[4] A few elastomers can hold large amounts of plasticizer and filler without appreciable degradation of properties, e.g., EPDM. 一些弹性体可以填充大量增塑剂和填料而不会导致性能的明显降低，例如EPDM。

Exercises

1. Translate the third paragraph of the text into Chinese.
2. Put the following words and phrases into English.

增塑剂	可塑性	流动性	混炼	压延
挤出	成型	回弹性	抗裂纹增长性	阻燃性能

3. What is the ASTM definition of plasticizer?

[Reading Material]

Introduction of Processing Aids

Processing aids are added to rubber compounds to help in process, e.g., mixing, calendering, extrusion and molding by reducing the viscosity.

Most commonly used processing aids are hydrocarbon oils. Such oils are classified into three groups: aromatic, naphthenic and paraffinic. Oils don't chemically react with rubber molecules but simply "lubricate" molecules, lower the viscosity and, thus, improve processability while reducing compound cost. Use of oils in rubber compounds permits application of higher molecular weight elastomers for better properties.

Other processing aids used in rubber compounds are fatty acids, waxes, organic esters, and low molecular weight polymers.

Chemical peptizers also reduce polymer viscosity but by reducing molecular chain length. Peptizers react with free radicals formed during mixing and hence shorten the molecular length.

Words and Expressions

lubricate	[ˈluːbrɪkeɪt]	v.	加油润滑,使润滑
processability	[prəusesəˈbɪlɪtɪ]	n.	成型性能,加工性能
fatty	[ˈfæti]	adj.	脂肪的
fatty acids			脂肪酸
wax	[wæks]	n.	石蜡
organic ester			有机酯
peptizer	[ˈpeptaɪzər]	n.	塑解剂
free radicals			自由基

Lesson 21 Antidegradants

Good aging properties of rubber compounds are essential for providing acceptable service life. [1] The type of elastomer used is the principal factor in considering aging properties. In general, the more saturated the main chain of elastomer, the better are the aging properties. For example, EPDM and butyl rubbers are quite stable as compared to more unstaturated polymers such as polyisoprene, SBR, and polybutadiene. Unsaturation sites are susceptible to oxidation and ozonization effects. Chemical antioxidants and antiozonants are uses to extend the service life of products containing diene rubbers.

Oxidation—[2] Oxidation proceeds by free radical mechanisms and leads to chain scission and crosslinking. In chain scission, free radicals attack the polymer backbone causing degradation. This happens mainly in natural rubber and butyl rubber oxidation. For SBR, EPDM, neoprene and nitrile rubber, free radicals formed during aging lead to degradation as well as crosslinking and hence hardening.

Heat, UV light and presence of transition metal ions (especially in natural rubber) accelerate the oxidation reaction. Flexing also accelerates the aging process. [3] Flexing involves not only a mechanical fatigue but also generates heat which accelerates the oxidation. Because oxidation is a chemical reaction, an increase of about 10℃ in temperature, almost doubles the rate of oxidation.

Antioxidants—Chemical antioxidants extend the life of rubber products by first reacting with polymeric free radicals and stopping propagation of polymer oxidation. Antioxidants are selected based on service requirements.

A good antioxidant should have low volatility, be stable at high temperatures and be soluble in rubber. Poor solubility will result in a bloom. For natural rubber compounds an antioxidant should provide protection against transition metal ions as well. In a non-staining, non-discoloring product, a non-staining antioxidant should be used.

Commercially available antioxidants generally fall in three groups: secondary amines, phenolics, and phosphites. Amine-type antioxidants are staining and are used in black compounds. The phenolics and phosphites are relatively non-staining and normally used in non-black or light-colored compounds.

Ozonization—[4] Ozone attack on unprotected elastomers takes place at the double bond via chain scission resulting in decomposition products such as ozonides, polymer peroxide, hydroperoxides, aldehydes, and ketones.

There are three different types of antiozonants: waxes, chemical antiozonants and ozone resistant elastomers with no or dangling unsaturation such as EPDM. [5] Waxes bloom to the surface, form film and provide protection under static conditions at relatively low cost. However, under dynamic conditions, surface wax film cracks providing sites for ozone attack. Thus waxes are not good antiozonants for dynamic applications of rubber.

<u>Chemical antiozonants</u>, such as DPPD, IPPD, and 6PPD diffuse to the surface and reacts with ozone before rubber molecules have chance to react with ozone. Some researchers verified this mechanism showing that ozonized antiozonant film is flexible and, thus, offers ozone protection even under dynamic conditions. 化学抗臭氧剂

[6] Therefore, a good antiozonant should be reactive with ozone, should have acceptable solubility and diffusion in rubber. Poor diffusion will slow down <u>migration</u> of the antiozonant to the surface and would offer poor protection. 迁移

Ozone tests are typically done at 30~50℃, at 10%~20% strain and 25~100 pphm ozone concentration. Tests are done under static as well as dynamic conditions.

New Words

aging	['eɪdʒɪŋ]	n.	老化
factor	['fæktə(r)]	n.	因素
susceptible	[sə'septəbl]	adj.	易受影响的
oxidation	[ˌɒksɪ'deɪʃn]	n.	氧化
ozonization	[ˌəʊzəʊnaɪ'zeɪʃn]	n.	臭氧化作用
anti-	['ænti]	n.& adj.	反,抗
antioxidant	[ˌænti:ɒk'sɪdænt]	n.	抗氧剂
antiozonant	[əntɪə'zəʊnənt]	n.	抗臭氧剂
proceed	[prə'si:d]	vi.	进行
scission	['sɪʒən]	n.	断裂
hardening	['hɑ:dnɪŋ]	v.	(使)变硬
flexing	['fleksɪŋ]	n.	屈挠,可挠性
volatility	[ˌvɒlə'tɪləti]	n.	挥发性
phenolic	[fɪ'nɒlɪk]	adj.	酚的
phosphite	['fɒsfaɪt]	n.	亚磷酸盐
ozonide	['əʊzəʊnaɪd]	n.	臭氧化物
hydroperoxide	[haɪdrəpə'rɒksaɪd]	n.	氢过氧化物
aldehyde	['ældɪhaɪd]	n.	醛
ketone	['ki:təʊn]	n.	酮
solubility	[ˌsɒljʊ'bɪləti]	n.	溶解度
diffusion	[dɪ'fju:ʒn]	n.	扩散
migration	[maɪ'greɪʃn]	n.	迁移
concentration	[ˌkɒnsn'treɪʃn]	n.	浓度

Notes

[1] The type of elastomer used is the principal factor in considering aging properties. In general, the more saturated the main chain of elastomer are, the better the aging properties are. 弹性体的种类是影响老化性能的主要因素。一般来说，弹性体主链饱和度越高，老化性能越好。

[2] Oxidation proceeds by free radical mechanisms and leads to chain scission and crosslinking. In chain scission, free radicals attack the polymer backbone causing degradation. （橡胶）氧老化属于自由基自催化反应机理，氧老化会导致（橡胶）分子链断裂和交联。在（橡胶）分子链断裂过程中，自由基进攻高分子主链导致降解。

[3] Flexing involves not only a mechanical fatigue but also generates heat which accelerates the oxidation. Because oxidation is a chemical reaction, an increase of about 10℃ in temperature, almost doubles the rate of oxidation. 屈挠不仅引起机械疲劳，还会产生热量，热量会加速氧老化。因为氧老化是一种化学反应，温度每上升约10℃，几乎导致氧老化速度加倍。

[4] Ozone attack on unprotected elastomers takes place at the double bond via chain scission resulting in decomposition products such as ozonides, polymer peroxide, hydroperoxides, aldehydes, and ketones. 臭氧通过进攻未受保护的弹性体的双键，导致链断裂并分解产生如臭氧化物、高分子过氧化物、氢过氧化物、醛和酮等物质。

[5] Waxes bloom to the surface, form film and provide protection under static conditions at relatively low cost. However, under dynamic conditions, surface wax film cracks providing sites for ozone attack. 石蜡喷出至（橡胶制品）表面，形成蜡膜，并以相对较低的成本在静态条件下对（橡胶制品）提供保护。然而，在动态条件下，表面蜡膜破裂，臭氧能进攻（橡胶制品）某些位点。

[6] Therefore, a good antiozonant should be reactive with ozone, should have acceptable solubility and diffusion in rubber. Poor diffusion will slow down migration of the antiozonant to the surface and would offer poor protection. 因此，好的抗臭氧剂应能与臭氧发生反应，且应在橡胶中具有适当的溶解度和良好的分散性。分散性不良会减缓抗臭氧剂向表面的迁移，导致抗臭氧效果差。

Exercises

1. Translate the fourth, fifth and sixth paragraphs of the text into Chinese.
2. Put the following words and phrases into English.

抗降解剂	老化	老化性能	主链	不饱和位点
氧老化	臭氧老化	紫外光老化	自由基机理	使用寿命
抗臭氧剂	溶解度	扩散	浓度	

[Reading Material]

Procedures for Compound Development for a Rubber Product

To develop a rubber product compound, it is good to first find out what formulations are being used by manufactures of that product. Since no one will reveal that information, the best way to find out is to analyze that commercial product and reconstruct the formulation based on the analysis. This becomes the starting point. The following sequence can be followed for compound development.

1. Determine current formulation (s) used.
2. Test rubber properties of competitors' products (if possible).
3. Define which properties need to be improved.
4. Determine if the reconstructed formulation used correct elastomers, type and amount of carbon black, type and amount of oil, types of curing agents, and antidegradants to meet the desired properties.
5. If not, then make changes in compounding ingredients and come up with several alternate formulations. One can use the design of experiments (DOE) to develop an optimum compound.
6. Mix, cure samples and determine properties of those compounds in laboratory.
7. Select one or two best compounds that have improved properties.
8. Run a plant trail for mixing and other processing steps and curing the product.
9. Test factory mixed compounds in laboratory for desired properties.
10. Test the actual product in step 8 above and determine if it meets the requirements.
11. Select the best compound and start production of products using that compound.
12. If steps 8～11 don't work out, go to step 5 and start all over again.

Words and Expressions

commercial	[kəˈmɜːʃl]	*adj.*	商业的，贸易的
reconstruct	[ˌriːkənˈstrʌkt]	*v.*	改造
competitor	[kəmˈpetɪtə(r)]	*n.*	竞争者，对手
define	[dɪˈfaɪn]	*v.*	规定，使明确
improve	[ɪmˈpruːv]	*v.*	改善，改良
design of experiments (DOE)			实验设计
plant trail			工厂小试

PART 4

Processing Methods in the Rubber Industry

Lesson 22 Introduction of Rubber Processing

Raw rubber as received from the manufacture, or from the plantation in the case of natural rubber, has few uses as such. It has to be mixed with various compounding ingredients, shaped, and vulcanized to give a usable end product. These three components, mixing, shaping, and curing, together, constitute rubber processing.

Mixing is central to rubber processing. If the base compound is inadequately mixed, problems cascade down through the subsequent processes or shaping and curing into the end product. There are three general shaping processes: extrusion, calendering and molding—compression, transfer, or injection type.

[1] Rubber is a difficult material to process because it has both viscous and elastic properties. But, even more than this, because of the infinite variety of possible compounds that can be prepared from any grade of rubber, it is difficult to predict mixing, molding, or extrusion behavior of a compound based on the properties of the raw rubber alone. [2] Depending on the type and quantity of carbon black, plasticizer, other fillers, etc. used, the resulting compound is formulated to meet end-application requirements, not merely to process well. The processing plant has to adjust its process to suit the compound, not the other way around. However, it is of no value to present the plant with a compound impossible to mix or process. Therefore, it has to be integration between the compound, process and product, involving extensive cooperation between compounder, mixer, and processor.

配合剂/成型

挤出/压延/成型模压/转移/注射
黏性的
弹性的

New Words

manufacture	[ˌmænjuˈfæktʃə(r)]	n.	工厂
inadequately	[ɪnˈædɪkwətlɪ]	adv.	不充分地

cascade	[kæˈskeɪd]	v.	串级
subsequent	[ˈsʌbsɪkwənt]	adj.	后来的，随后的
extrusion	[ɪkˈstruːʒn]	n.	挤出
calendering	[kæˈlɪndərɪŋ]	n.	压延
molding	[ˈməʊldɪŋ]	n.	成型
compression	[kəmˈpreʃn]	n.	模压
transfer	[trænsˈfɜː(r)]	n.	转移
injection	[ɪnˈdʒekʃn]	n.	注射
viscous	[ˈvɪskəs]	adj.	黏的，黏性的
infinite	[ˈɪnfɪnət]	adj.	无限的
formulate	[ˈfɔːmjuleɪt]	v.	构想出，规划
merely	[ˈmɪəli]	adv.	仅仅，只不过
extensive	[ɪkˈstensɪv]	adj.	广泛的

Notes

[1] Rubber is a difficult material to process because it has both viscous and elastic properties. But, even more than this, because of the infinite variety of possible compounds that can be prepared from any grade of rubber, it is difficult to predict mixing, molding, or extrusion behavior of a compound based on the properties of the raw rubber alone. 橡胶是一种难以加工的材料，因为它具有黏弹性。此外，由于可以从任何等级的橡胶制备出无数种可能的胶料，基于生胶的性质难以预测胶料的混炼、成型或挤出工艺性能。

[2] Depending on the type and quantity of carbon black, plasticizer, other fillers, etc. used, the resulting compound is formulated to meet end-application requirements, not merely to process well. 根据所使用的炭黑、增塑剂、其他填料等的种类和数量，所制得的胶料需满足最终应用的要求，而不仅仅是加工性良好。

Exercises

1. Translate the following two paragraphs into Chinese.

The rubber technologist's mixing department has bags of powders, drums of liquids and bales or granules or chips of raw gum elastomer. These are weighed out precisely, to match both the batch weight needed and the ratio of ingredients in the formulation. Machines are necessary to mix these chemicals, resulting in a finely blended, solid homogeneous mixture. In many cases, the compounder and process operator expend their energy reducing the elastic component of the uncured rubber compound, to help it process, and then increase that component again during vulcanization.

Mixing is accomplished using mills and/or internal mixing machines. The resulting compound is then preshaped by mills, extruders or calenders, to prepare it for vulcanization. The latter is achieved using molds (which further shape the product), autoclaves, and sometimes ovens. That just leaves finishing operations, such as removing flash, or maybe the grinding of rubber rollers (cured in an autoclave) to a finished dimension, and then packaging the product.

2. Put the following words and phrases into English.

| 混炼 | 挤出 | 压延 | 成型 | 模压成型 |
| 转移成型 | 注射成型 | 工艺性能 | 黏性 | 弹性 |

3. What are the three general shaping processes?

[Reading Material]

Introduction of Rubber Mastication

The production of rubber goods involves two basic operations—compounding and curing. Rubber polymers behave as viscoelastic fluids when sheared at elevated temperatures. This enables the incorporation of the various fillers and chemical additives in the process of compounding.

The first step in rubber compounding is mastication or polymer "breakdown". This is essentially development of the polymer's viscoelasticity to make it receptive to the additives. Most synthetic elastomers are produced with the uniformity in chemistry, viscosity and stability that minimizes or precludes the need for mastication. This step is, nevertheless, usually necessary for compounds containing a blend of polymers to provide a uniform mixture prior to further compounding. Natural rubber is characteristically widely variable in viscosity from lot to lot, so that several lots are usually blended and masticated to control viscosity and physical properties of the compound. The breakdown of natural rubber promotes viscoelasticity by extending or disentangling the polymer chains or by severing the chains. Breakdown cycles for both natural and synthetic rubber are often facilitated by the inclusion of chemical plasticizers. Polymer breakdown and subsequent compounding was historically accomplished on open roll mills, although today these operations are almost always performed in an internal mixer. The open mill consists of two metal rolls which are jacketed for temperature control. These rolls turn toward each other at fixed separations and often at different speeds. This provides high shear mixing forces. Banbury mixers are the most commonly used internal mixers today. They consist of two rotor blades turning toward each other in an enclosed metal cavity. Mixing shear, and consequent mixing time, is determined by rotor shape, size, and speed. A large Banbury can produce 500kg of finished compound in a matter of minutes.

Words and Expressions

mastication	[ˌmæstɪˈkeɪʃn]	n.	塑炼
viscoelastic	[ˌvɪskʊʊˈlæstɪk]	adj.	黏弹性的
viscoelastic fluids			黏弹性流体
viscoelasticity	[ˈvɪskəʊlæsˈtɪsɪtɪ]	n.	黏弹性
incorporation	[ɪnˌkɔːpəˈreɪʃn]	n.	配合
compounding	[ˈkɒmpaʊndɪŋ]	n.	配合
receptive	[rɪˈseptɪv]	adj.	能容纳的
preclude	[prɪˈkluːd]	v.	阻止,排除
disentangling	[ˌdɪsenˈtæŋɡəlɪŋ]	v.	解开…的结
chemical plasticizer			化学增塑剂

open roll mill			开炼机
internal mixer			密炼机
separation	[ˌsepəˈreɪʃn]	n.	间距，间隔
shear mixing force			剪切共混力
blade	[bleɪd]	n.	桨叶
cavity	[ˈkævəti]	n.	腔，模腔
in a matter of minutes			几分钟之内

Lesson 23 Rubber Mixing

Overview of Mixing

Mixing can be subdivided into three steps:

1. Feeding the ingredients to the mixer in the correct quantities, at the correct times, and at the correct temperatures.

2. The actual mixing of the ingredients.

3. Discharge of the mixed compound from the mixer, and its shaping, cooling, and packing for the next process.

[1] The single most important aspect of mixing is consistency: consistency in type, quality, and quantity of ingredients; consistency in temperatures, speeds, and pressure; consistency in ingredient addition times, mixing stage times, and dumping times; consistency in work input, shear rates, and shear stress.

[2] Only if all these things are invariable, batch after batch, will the product be consistent in its behavior in the subsequent processes of shaping and curing, and as a result have consistent finished product properties.

One should differentiate between compounding and mixing. Compounding is deciding, which base elastomer, or elastomers, together with the quantities and types of other ingredients are required to give a product which, after mixing and curing, will have the properties required for the specific application. In a well-designed compound, each ingredient and level are selected to achieve specific properties in that compound. Once the compound is in place, design of the mixing process involves deciding what equipment to use, and the appropriate speeds, pressures, times, temperatures and procedures required to blend those ingredients into an adequately mixed compound. In other words, mixing is a multifaceted process.

Mixing Operation

(1) **Operations overview**—[3] The aim of the rubber mixing process is to produce a product that has the ingredients dispersed and distributed sufficiently thoroughly to permit it to shape readily, cure efficiently, and give the required properties for the application, all with the minimum expenditure of machine time and energy.

There are four main processes involved in the mixing operation. These are incorporation, dispersion, distribution, and plasticization (viscosity reduction).

Incorporation is the first stage of mixing, during which the compounding ingredients form a coherent mass. Dispersion is the process during which the filler agglomerates are reduced to their ultimate size and dispersed in the rubber. Distribution is simple homogenization, during which the various ingredients are randomly distributed throughout the mass of the mix. During plasticization, the mix reaches its final viscosity as plasticizers effectively lubricate the mix.

[4] These four processes are not entirely distinct, they all take place throughout the mixing cycle, but incorporation predominates in the early stages, dispersion in the middle, and plasticization toward the end.

It is important to avoid over-mixing. It wastes time and energy and can turn a profitable operation into an unprofitable one. In addition, exposure to shearing at high temperatures can result in excessive interaction between carbon black and rubber, crosslinking and viscosity increases, and, in the case of natural rubber, and some synthetics, breakdown of the rubber itself.

(2) *Temperature control*—This is a major concern in any mixing operation. The energy that produced the shearing action is largely converted into heat, and this results in temperature rise during mixing. [5] Rubber is an inherently poor conductor of heat, and heat can be removed from the mass of rubber only if fresh surfaces of rubber are generated and brought into contact with cooler metal surfaces—that is, the surface of the rotors and the inside of the chamber of internal mixer or the roller of the open two-roll mill.

(3) *Mixing procedures*—There are five main types of ingredients in a normal rubber mix: rubber, fillers, plasticizers, antidegradants, and curatives. In most mixing operations, curatives are added at a second, lower temperature stage, either on a mill or in a high shear mixer or a relatively low shear mixer.

The choice of material input sequence to a mixer profoundly affects the efficiency of mixing. [6] In general the number of additions in the mixing cycle should be minimized because each addition requires the ram to be raised, and with no pressure on the mix, little effective mixing occurs.

(4) *Mill mixing*—Open mills were the original equipment used in mixing rubber compounds and are still used for specialty compounds, for small batches, in small shops where the production volume does not warrant expenditure on an internal mixer, and for curative addition in a second pass.

Temperature control is very important in mill mixing. [7] Cooling is usually accomplished by flooding or spraying the inside of the mill rolls with water, or by circulating water through channels drilled in the roll walls. Drilled rolls provide the best heat exchange, and thus the best temperature control. Compound temperature is adjusted by regulating the rate of flow and temperature of the water through the rolls.

New Words

subdivide	[ˌsʌbdɪˈvaɪd]	v.	再分,细分
mixer	[ˈmɪksə(r)]	n.	混炼机
discharge	[dɪsˈtʃɑːdʒ]	v.	放出;流出
consistency	[kənˈsɪstənsi]	n.	协调一致性
dumping	[ˈdʌmpɪŋ]	n.	排料;倾倒
invariable	[ɪnˈveəriəbl]	adj.	恒定的,始终如一的
differentiate	[ˌdɪfəˈrenʃieɪt]	v.	区分,辨别
adequately	[ˈædɪkwətlɪ]	adv.	足够地;适当地
multifaceted	[ˌmʌltiˈfæsɪtɪd]	adj.	多方面的,多才多艺的
sufficiently	[səˈfɪʃntlɪ]	adv.	足够地,充分地
thoroughly	[ˈθʌrəli]	adv.	彻底地
expenditure	[ɪkˈspendɪtʃə(r)]	n.	(材料等的)耗费
coherent	[kəʊˈhɪərənt]	adj.	一致的
agglomerate	[əɡˈlɒməreɪts]	n.	聚集体
disperse	[dɪˈspɜːs]	v.	(使)分散,使粒子分散
homogenization	[həʊmədʒənaɪˈzeɪʃn]	n.	均化作用,均匀化
lubricate	[ˈluːbrɪkeɪt]	v.	加油润滑,使润滑
predominate	[prɪˈdɒmɪneɪt]	v.	在…中占优势,支配
inherently	[ɪnˈhɪərəntlɪ]	adv.	天性地,固有地
exposure	[ɪkˈspəʊʒə(r)]	n.	暴露
rotor	[ˈrəʊtə(r)]	n.	转子
chamber	[ˈtʃeɪmbə(r)]	n.	模腔
roller	[ˈrəʊlə(r)]	n.	辊筒
curative	[ˈkjʊərətɪv]	n.	硫化剂
efficiency	[ɪˈfɪʃnsi]	n.	效率
ram	[ræm]	n.	柱塞
specialty	[ˈspeʃəlti]	adj.	特色的;专门的
expenditure	[ɪkˈspendɪtʃə(r)]	n.	费用;支出
flooding	[ˈflʌdɪŋ]	n.	溢流
spray	[ˈspreɪ]	v.	喷

Notes

[1] The single most important aspect of mixing is consistency: consistency in type, quality, and quantity of ingredients; consistency in temperatures, speeds, and pressure; consistency in ingredient addition times, mixing stage times, and dumping times; consistency in work input, shear rates, and shear stress. 混炼工艺中最重要的是协调一致性:配合剂的种类、质量和数量

的协调一致性；混炼温度、速度和压力的协调一致性；配合剂添加时间、各阶段混炼时间、排料时间的协调一致性；功率输入、剪切速率和剪切应力的协调一致性。

[2] Only if all these things are invariable, batch after batch, will the product be consistent in its behavior in the subsequent processes of shaping and curing, and as a result have consistent finished product properties. 只有当一批批胶料的混炼工艺条件都保持不变时，产品在后续成型和硫化过程中的工艺性能才会一致，并因此具有一致的成品性能。

[3] The aim of the rubber mixing process is to produce a product that has the ingredients dispersed and distributed sufficiently thoroughly to permit it to shape readily, cure efficiently, and give the required properties for the application, all with the minimum expenditure of machine time and energy. 橡胶混炼工艺过程的目的是在最短的混炼时间和能量消耗最少的条件下制备出一种配合剂分散充分且均匀的胶料，以使得胶料易于成型、硫化，且性能符合（橡胶制品）应用的需求。

[4] These four processes are not entirely distinct, they all take place throughout the mixing cycle, but incorporation predominates in the early stages, dispersion in the middle, and plasticization toward the end. 这四个工艺过程并不严格相互独立，它们存在于整个混炼的循环过程中，但是配合主要发生在混炼早期阶段，分散主要发生在中间阶段，增塑主要发生在后期阶段。

[5] Rubber is an inherently poor conductor of heat, and heat can be removed from the mass of rubber only if fresh surfaces of rubber are generated and brought into contact with cooler metal surfaces—that is, the surface of the rotors and the inside of the chamber of internal mixer or the roller of the open two-roll mill. 橡胶本身是一种热的不良导体，只有当混炼过程中橡胶产生新的表面并与较冷的金属表面——即转子表面和密炼机模腔内表面或开炼机辊筒接触时，热量才能从橡胶基体中去除。

[6] In general the number of additions in the mixing cycle should be minimized because each addition requires the ram to be raised, and with no pressure on the mix, little effective mixing occurs. 一般而言，在混炼周期中，应使得加料次数最少化，因为每次加料都需要提起柱塞，导致没有压力施加到混炼胶料上，混炼效果减弱。

[7] Cooling is usually accomplished by flooding or spraying the inside of the mill rolls with water, or by circulating water through channels drilled in the roll walls. 冷却通常是通过在开炼机辊筒内灌注或喷洒水，或通过在辊筒内壁上安装管道循环水来完成的。

Exercises

1. Translate the following two paragraphs into Chinese.

As a compounding process, masterbatching means incorporation of the other compounding ingredients into the rubber, except for the curing system. By omitting the cure system, the masterbatch may be intensively mixed without fear that the high temperatures generated will cause premature vulcanization. The objectives of masterbatching are homogeneous blending of polymer and chemical additives, the best possible de-agglomeration and dispersion of fillers, and the development of the proper final viscosity. This viscosity, in turn, is a function of the homogeneity achieved in the mixing of the polymer and additives. Because considerable heat is generated during milling, the masterbatch operation must often be terminated before its objectives are attained. This is to avoid heat degradation. In such cases, the compound is remilled before addition of the curing system. Masterbatches are typically prepared in a Banbury mixer, after which they are discharged and cooled.

Separate from the compounding process, masterbatches are often prepared of individual additives that are difficult to disperse during compounding, zinc oxide for example. A concentrated dispersion of the additives in polymer is preformed. This can then be added to the rubber compound for faster incorporation of the additive because it is essentially predispersed.

2. Put the following words and phrases into English.

混炼机	双辊开炼机	密炼机	转子	模腔
柱塞	辊筒	加料顺序	排料时间	温度控制
配合	分散	增塑	过炼	剪切作用
聚集体	母炼胶	辊距	辊缝	进料口

3. What is the difference between compounding and mixing?
4. What is the aim of the rubber mixing process?
5. Describe the specific procedure of incorporation.

[Reading Material]

Introduction of Open Mills and Internal Mixing Machines

Open Mills—Open mills were used at the beginning of the rubber industry and are still an essential piece of rubber processing equipment. Open two-roll mill consists of two horizontally placed hollow metal cylinders rotating towards each other (see Figure 4.1). The distance between the cylinders can be varied, typically between 0.25 to 2.0cm. This gap between the rolls is called a nip.

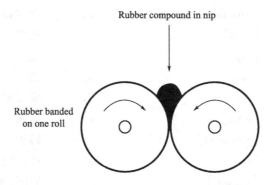

Figure 4.1 Conceptual view of rubber mill rolls

Internal mixing machines—If the rolls of a mill are twisted to produce a corkscrew effect (they would now be called rotors), and then a block of steel is placed over the mill nip with the block connected to a steel rod above it, this would be called a ram. The ram would move up, to allow addition of ingredients to the nip, and it would move down to force the compound ingredients into the nip. If the whole thing is surrounded in a heavy metal jacket with a chute at the top to put ingredients in and a door at the bottom (underneath the rotors), to let the mixed material out, the result will be an internal mixing machine (see Figure 4.2).

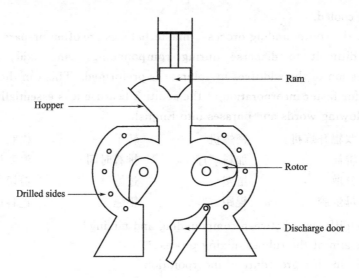

Figure 4.2　Conceptual cross-section through an internal mixing machine, with tangential rotors

Words and Expressions

masterbatching			母炼胶的制备
open mill			开炼机
open two-roll mill			双辊开炼机
conceptual	[kən'septʃuəl]	adj.	观念的,概念的
horizontally	[ˌhɒrɪ'zɒntəlɪ]	adv.	水平地,横地
hollow	['hɒləʊ]	adj.	空的,中空的
cylinder	['sɪlɪndə(r)]	n.	圆柱,圆筒,圆柱体
gap	[gæp]	n.	缝隙,辊隙
nip	[nɪp]	n.	辊距
twiste	[twɪst]	v.	捻,绕
corkscrew	['kɔːkskruː]	adj.	螺旋形的
block	[blɒk]	n.	活塞
rotor	['rəʊtə(r)]	n.	转子
rod	[rɒd]	n.	杆,拉杆
bottom	['bɒtəm]	n.	底部,末端
underneath	[ˌʌndə'niːθ]	prep.	在…下面
internal mixing machine			密炼机
cross-section	['krɒs'sekʃən]	n.	横断面;横截面图
hopper	['hɒpə(r)]	n.	进料口
discharge door			排料口

Lesson 24 Rubber Extrusion

Overview of Extrusion

Extrusion of profiles, tubing, and hose is one of the most important fabrication processes in the rubber industry. The extruder itself has two main functions: to pump the rubber compound through the barrel; and to generate enough pressure in the process to force the material through a die to give the required cross-sectional shape. [1] The compound, usually in the form of a strip, is fed at a controlled rate into the extruder, which imparts heat, moves the material along the machine, and generates the necessary pressure. After leaving the die the product has to be cured, cooled, often cut to length, and packaged for transport to the customer.

Extruder Operation and Control

The three contributors to extrusion variability are: changes in the feedstock; inconsistent feed flow; and the extruder behavior itself.

Consistency in feeding rate is crucial to the uniformity of extruder output. The feed throat must have a rolling bank of consistent size, because if it fluctuates the heat input per unit volume will change, and therefore the temperature of the feed entering the extruder will change. This ultimately affects output rate and extrude dimensions.

Two potential sources of variation in the extruder are screw speed and zone temperatures. [2] Speed variations are obviously reflected in output variations. As long as there is no mechanical problem in the drive, modern feedback controlled dc drives have more than adequate capacity to hold the speed constant. Equally, control of the zone temperatures cooling medium can also be very precise.

Extrudate temperature cannot be allowed to exceed the threshold temperature, above which significant crosslinking starts to occur. [3] At very low screw speeds, the long residence time of the rubber in the extruder allows heat conduction from the metal components in contact with the rubber to be the main factor in determining the extrudate temperature. Residence time should be long enough to achieve good thermal homogeneity in the extrudate.

[4] As screw speed rises, the amount of mechanical energy dissipated per unit mass of rubber increases, until the temperature of the stock exceeds the set zone temperatures and the direction of heat flow reverses. At high speeds, especially with large machines, conduction to the barrel ceases to have any significant effect on the stock temperature and any further rise in screw speed produces scorched extrudate.

[5] Heat from mechanical shear is not developed uniformly in the stock, as speed is increased, and conduction becomes less effective in achieving thermal homogeneity. This is why screws have mixing sections,

	橡胶挤出
	型材/管材/胶管
	挤出机
	胶料/机筒
	口模（型）
	横截面的/条状形式
	硫化/冷却
	切成一定长度/包装
	原料
	喂料速度的稳定
	输出/喂料口
	每单位体积的热量输入
	输出速率/挤出尺寸
	螺杆转速
	区域温度
	反馈控制直流驱动器
	区域温度冷却介质
	挤出温度/临界温度
	停留时间
	热传导
	主要因素
	热均匀性
	机械能耗
	每单位质量橡胶
	热流
	焦烧挤出物
	机械剪切
	热均匀性/混合组件

to mix and distribute the heat generated.

　　Screw and barrel temperatures have a significant effect on extruder performance. [6]Although the <u>bulk temperature</u> of the stock is not determined by the <u>set temperatures</u> of the screw or the barrel zones, the rubber actually in contact with the metal does experience those temperatures. Thus the <u>adhesion</u> between the rubber and the screw and barrel does <u>depend on</u> the set temperatures. Cold rubber will not adhere to a cold metal surface. At some temperature depending on the type and <u>grade of rubber</u>, but usually in the range of 50~70℃ adhesion between rubber and metal is developed. Above that temperature adhesion decreases again as the rubber softens. Thus for <u>maximum output</u>, the feed zone should be at that temperature, and the screw temperature should remain high with the barrel kept at a lower temperature.

整体温度
设定温度

黏附力
取决于
橡胶等级

最大输出

New Words

profile	['prəʊfaɪl]	n.	型材；侧面
tubing	['tjuːbɪŋ]	n.	管材；管状材料
hose	[həʊz]	n.	软管，胶皮管
fabrication	[ˌfæbrɪ'keɪʃn]	n.	制造
extrusion	[ɪk'struːʒn]	n.	挤出
extruder	[eks'truːdə]	n.	挤出机
pump	[pʌmp]	v.	用泵输送
barrel	['bærəl]	n.	机筒
die	[daɪ]	n.	口模，口型
cross-section	['krɒs'sekʃən]	n.	横截面；横截面图
strip	[strɪp]	n.	带状；长条
variability	[ˌveərɪə'bɪlətɪ]	n.	变化性
feedstock	['fiːdstɒk]	n.	给料；进料
inconsistent	[ˌɪnkən'sɪstənt]	adj.	不一致的；前后矛盾的
uniformity	[ˌjuːnɪ'fɔːmətɪ]	n.	均匀性
output	['aʊtpʊt]	n.	产量；输出
screw	[skruː]	n.	螺杆
fluctuate	['flʌktʃueɪt]	v.	波动
throat	[θrəʊt]	n.	咽喉；管颈
zone	[zəʊn]	n.	区域；地带
capacity	[kə'pæsətɪ]	n.	生产能力
threshold	['θreʃhəʊld]	adj.	阈值的，临界值的
homogeneity	[ˌhɒmədʒə'niːətɪ]	n.	一致性；均匀性
dissipate	['dɪsɪpeɪt]	v.	消耗，耗散
adhesion	[əd'hiːʒn]	n.	黏合力；黏附力

cease	[siːs]	v.	终止,停止
scorch	[skɔːtʃ]	v.	焦烧
bulk	[bʌlk]	n.	主体

Notes

[1] The compound, usually in the form of a strip, is fed at a controlled rate into the extruder, which imparts heat, moves the material along the machine, and generates the necessary pressure. 将胶料（通常为条形）按一定的速度喂入挤出机中，该挤出机赋予（胶料）热量，使胶料沿着挤出机向前移动并产生必要的压力。

[2] Speed variations are obviously reflected in output variations. As long as there is no mechanical problem in the drive, modern feedback controlled dc drives have more than adequate capacity to hold the speed constant. 螺杆转速的变化在输出速率变化中有显著反映。只要驱动器没有机械问题，现代的反馈控制直流驱动器就足以保持（螺杆）转速的稳定。

[3] At very low screw speeds, the long residence time of the rubber in the extruder allows heat conduction from the metal components in contact with the rubber to be the main factor in determining the extrudate temperature. Residence time should be long enough to achieve good thermal homogeneity in the extrudate. 在非常低的螺杆转速下，橡胶在挤出机中具有较长的停留时间，致使与橡胶接触的金属组件的热传导成为决定挤出物温度的主要因素。停留时间应该足够长到可以保证挤出胶料获得良好的热均匀性。

[4] As screw speed rises, the amount of mechanical energy dissipated per unit mass of rubber increases, until the temperature of the stock exceeds the set zone temperatures and the direction of heat flow reverses. 随着螺杆转速的升高，每单位质量橡胶的机械能消耗量增加，直到胶料温度超过区域设定温度并且热流方向逆转。

[5] Heat from mechanical shear is not developed uniformly in the stock, as speed is increased, and conduction becomes less effective in achieving thermal homogeneity. This is why screws have mixing sections, to mix and distribute the heat generated. 随着螺杆转速的增加，胶料中由机械剪切产生的热量不均匀，并且热传导对实现热均匀性的效果变差。这就是螺杆具有混合组件的原因，以混合和分散所产生的热量。

[6] Although the bulk temperature of the stock is not determined by the set temperatures of the screw or the barrel zones, the rubber actually in contact with the metal does experience those temperatures. Thus the adhesion between the rubber and the screw and barrel does depend on the set temperatures. 尽管胶料的主体温度不是由螺杆或机筒区域的设定温度决定的，但实际上与金属（螺杆和机筒）接触的橡胶确实受到其温度的影响。因此，橡胶与螺杆和机筒之间的黏合力取决于该设定温度。

Exercises

1. Translate the following two paragraphs into Chinese.

Screw-type extruders are used both in the compounding operations and to form articles for vulcanization. The extruder screw will force the rubber through a die to form pellets, or through a pair of rolls to form a sheet. In either case, the extrusion operation allows the compound to cool down while it is turned into a more easily handled form.

Heated mixing extruders can be fitted to improve compound uniformity and reduce viscosity, and they can warm up material for calender feed. Rubber compound in strip or pellet form is fed to the extruder where it is heated and extruded in slug or rope form into the nip of a rubber calender. This type of extruder can also be used to take cold finished compound in strip, pellet, or slab form, heat it, blend it, and extrude it in the desired shape. This can be a finished shape, such as a rubber hose, which can be continuously vulcanized right after the extruder. This can also be a component for fabrication with other parts into a more complex item such as a tire. Where the shape must be very precise, the compound may be first warmed on a mill or warm-up extruder and then fed to a shaping extruder.

2. Put the following words and phrases into English.

挤出	挤出机	橡胶挤出	机筒	口型
喂料口	螺杆	型材	管材	胶管
螺杆转速	临界温度	停留时间	冷却介质	热均匀性

[Reading Material]

Introduction of Extruder and Die Swell

Introduction of Extruder

Extruders (see Figure 4.3) are conceptually a pump, consisting of a screw to move the material forwards, a barrel around the screw to contain the material, help it move, and provide part of the temperature control. The back end has a hopper, sometimes with feed rollers, to put rubber into the screw, and the front end has a "head" to hold a die, through which the rubber extrudes.

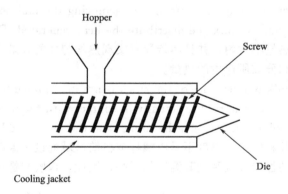

Figure 4.3 Conceptual view of a basic extruder

The extruder system is designed to build up compression at the discharge end, to ensure consolidation of the material in the head. This can be realized in a number of ways, such as reducing the screw pitch towards the front. An important design variable is the ratio of the length to the diameter of the screw, the L/D ratio.

Extruders that use pre-warmed rubber compound, hot feed extruders, (pre-warming on a mill for example) use a small ratio, for example 6∶1, while those using rubber compound at room temperature, cold feed extruders, need a larger ratio, for example 12∶1. This longer length is needed since the initial part of the screw is used to heat up the compound. Some extruders have a vent from the screw cavity through the barrel to the outside, to allow the escape of any air trapped in the compound.

Die Swell

The die is designed to avoid sudden discontinuities, as the compound moves through it and thus often has a contoured lead (entrance) section. As the extrusion exits the die, the extrusion can shorten in length and increase in cross-section. This is known as die swell, which is dependent on die design, screw speed, temperature and the compound's viscosity and its elastic component.

In practice, die swell can be quite complex and it might be necessary to modify the die a number of times, before the required extrusion shape is achieved. This recognizes that even uncured rubber has complex elastic and plastic behavior. Like an elastic band it can undergo elastic recovery on exiting the die. The chemist tries to formulate his compound to decrease uncured elastic behavior and increase plasticity. For example, he might do this by adding a material called Factice or other process aids to the formulation. Increased plasticity can also be achieved if the temperature of the compound moving through the extruder is progressively increased as it moves towards the die and it then becomes softer (lower viscosity) and more plastic. A very low hardness vulcanizate is often made from a low viscosity uncured compound. This can sometimes cause sagging of the extrusion before it is cured. If some of the raw gum elastomer in the compound is replaced by a partially cross-linked grade, then the firmness of the extrudate is improved. This helps control the extrudate dimensions. Such elastomer grades are available for CR, NBR, NR and SBR.

Words and Expressions

die swell			挤出收缩膨胀
conceptually	[kən'septʃʊəlɪ]	*adv.*	概念地
pump	[pʌmp]	*n.*	泵
back end			后端
hopper	['hɒpə(r)]	*n.*	(漏斗形)储料器
roller	['rəʊlə(r)]	*n.*	辊筒
feed roller			进料辊
preform	['priː'fɔːmz]	*n.*	预制件
discharge end			出(排)料口
consolidation	[kən‚sɒlɪ'deɪʃən]	*n.*	巩固;变坚固
ratio of the length to the diameter			长径比
hot feed extruder			热喂料挤出机
cold feed extruder			冷喂料挤出机
vent	[vent]	*n.*	排气孔
discontinuity	[‚dɪs‚kɒntɪ'njuːətɪ]	*n.*	不连续,中断
contoured	['kɒntʊəd]	*adj.*	波状外形的
factice	['fæktɪs]	*n.*	(硫化)油膏,油胶
sagging	['sægɪŋ]	*n.*	下垂(沉,陷)

Lesson 25 Calendering of Rubber

Overview of Calendering

Calendering machines are used to produce continuous sheets from rubber compounds, sometimes for incorporating reinforcing materials such as textile or wire cord and for impregnating. A calender is similar to a mill in its operation, in that the rubber and other components pass between two or more metal rolls. However, in order to produce product of controlled thickness, significantly higher pressures are needed and therefore calender rolls, journals, and frames have to be more carefully designed and more sturdy than those of mills.

[1] The rolling gap formed in the nip during the calendering process readily entraps air. Thus, high nip compression is required to squeeze out this air. As a result standard calender lines cannot produce sheet free of air inclusions thicker than 2~3mm. Thicker sheets can only be produced by laminating thinner sheets, or by using a combination of a roller-die and calender, which can produce sheets up to 15mm.

Calendering Process

(1) **Feeding**— [2] To ensure steady operation of the calender, and to control shrinkage, the compound has to be preheated to around 200°F and fluxed before being fed to the calender. This is often done on an open mill. Alternatively, cold-feed pin-barrel extruders are increasingly used for the preheating step, as they require less energy per unit weight of compound to produce strip feed suitable for a calender.

[3] If a cord or fabric is also to be fed to the calender, the spools are fed from equipment which is pneumatically controlled to ensure constant tension in the feed. Accumulators are also required to allow constant feed to the calender compensating for splicing and feed spool changes.

(2) **Sheeting**—This process is normally carried out on a three-roll calender, with thickness control by a feedback system from the product. Usually, the rolls are crowned to compensate for deflection under load, and so to maintain a constant roll gap across the width of the sheet.

(3) **Frictioning**—This process impregnates a textile or metallic fabric between two rolls running at different speeds so that the rubber compound is forced into the interstices of the substrate.

(4) **Coating**—Coating differs from frictioning in that the rolls run at the same speed and, as a result, the compound is only laid down and pressed onto the substrate, at the required thickness.

(5) **Downstream processes**—An important requirement for calendered sheet is the measurement of thickness. This can be done using electromechanical sensors, beta gauges, or lasers. These control systems feedback adjustments to roll speed, and other roll controls as necessary.

Drum-cooling equipment is normally used to remove heat from the product, and is arranged so that the cooling takes place from both sides of the sheet. Pull-off speed can be adjusted to compensate for shrinkage as the product cools. Rubber sheets, which do not contain reinforcement, have to be supported on a belt as they are cooled.

药品冷却设备

牵引速度

Often, in order to prevent adhesion in storage, the surfaces are coated with zinc stearate, applied by brush rolls.

硬脂酸锌

New Words

calendering	[kæˈlɪndərɪŋ]	n.	压延
calender	[ˈkælɪndə]	n.	压延机(设备)
incorporate	[ɪnˈkɔːpəreɪt]	v.	使混合
textile	[ˈtekstaɪl]	n.	纺织品
impregnating	[ɪmˈpreɡneɪtɪŋ]	n.	浸渍
sturdy	[ˈstɜːdi]	adj.	坚固的,耐用的
entrap	[ɪnˈtræp]	v.	卷入;使陷入圈套
laminating	[ˈlæmɪneɪtɪŋ]	n.	层压(法)
feeding	[ˈfiːdɪŋ]	n.	喂料
shrinkage	[ˈʃrɪŋkɪdʒ]	n.	收缩
preheat	[ˌpriːˈhiːt]	v.	预热
flux	[flʌks]	v.	熔融
cord	[kɔːd]	n.	帘线
fabric	[ˈfæbrɪk]	n.	织物
spool	[spuːl]	n.	卷轴,线轴
pneumatically	[njuːˈmætɪkəli]	adv.	由空气作用
compensate	[ˈkɒmpenseɪt]	v.	补偿
compensating	[ˈkɒmpenseɪtɪŋ]	n.	补偿,修正
splicing	[ˈsplaɪsɪŋ]	n.	胶黏
deflection	[dɪˈflekʃn]	n.	偏斜,歪曲
frictioning	[ˈfrɪkʃənɪŋ]	n.	擦胶
metallic	[məˈtælɪk]	adj.	金属的
interstice	[ɪnˈtɜːstɪsɪz]	n.	空隙,缝隙
substrate	[ˈsʌbstreɪt]	n.	底物;基底
electromechanical	[ɪˈlektrəʊmɪˈkænɪkəl]	adj.	电动机械的
laser	[ˈleɪzə(r)]	n.	激光
adjustment	[əˈdʒʌstmənt]	n.	调解,调整
stearate	[ˈstiːəreɪt]	n.	硬脂酸盐

Notes

[1] The rolling gap formed in the nip during the calendering process readily entraps air.

Thus, high nip compression is required to squeeze out this air. 在压延过程中，辊距中形成的辊缝容易夹带空气。因此，需要通过高的辊筒压力来排出空气。

[2] To ensure steady operation of the calender, and to control shrinkage, the compound has to be preheated to around 200°F and fluxed before being fed to the calender. 为确保压延机的稳定运行和收缩的有效控制，必须将胶料预热至200°F左右，使胶料在喂入挤出机之前已处于熔融状态。$[t/℃ = \frac{5}{9}(t/℉ - 32)]$

[3] If a cord or fabric is also to be fed to the calender, the spools are fed from equipment which is pneumatically controlled to ensure constant tension in the feed. 如果要将帘线或织物喂入压延机，则由气动控制装置控制（帘线或织物）卷轴，以确保喂料时的恒定张力。

Exercises

1. Translate the following paragraph into Chinese.

Rubber calenders consist of at least three rolls which can be adjusted for gap, speed and temperature. They are used to form rubber sheeting to required lengths and thickness for subsequent operations. They are also used for frictioning or skim coating fabrics. Cord and wire are coated in this way for making plies used in the construction of tires and conveyor belts.

2. Put the following words and phrases into English.

压延	压延设备	压延生产线	压延工艺	三辊压延机
辊隙	辊筒	喂料	擦胶	贴胶
收缩	预热	帘线	织物	卷轴

3. Translate the first paragraph of the text into Chinese.

[Reading Material]

Introduction of Calenders

A calender can be crudely thought of as a very high precision mill, with the rolls stacked on top of one another, with anything from two to four rolls in various configurations. The distance between the rolls can be varied to control calendered thickness.

As with the extruder, the calender (see Figure 4.4) further processes the compound after it has been mixed in the internal mixer or on the mill. Calendering is a useful technique, if the final product is to be a roof or tank lining, expansion joint or indeed any further product which needs accurately dimensioned continuous sheet. Calendering is also used for applying rubber compound to textiles. Sheet from a mill will have a thickness which is much too imprecise, can have a rough surface and may contain some trapped "bubbles" of air. This makes it less desirable for processing into the above products. A more precise machine is preferred, and the calender fits this need.

A three-roll calender is very popular, where the middle roll is fixed, while the ones above and below it can be moved vertically to adjust the gap between the rolls. Calenders are extremely robust and solidly built machines, and may provide service for many decades. Some of the rolls can be a substantial size, i.e., 90cm in diameter and 250cm in length. The early calenders must

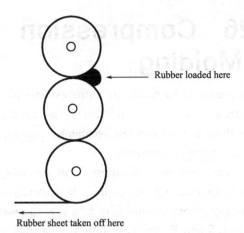

Figure 4.4 Conceptual end view of a basic, three roll vertical calender

have appeared as quite imposing pieces of machinery. Willshaw mentions "Chaffee's Monster" and "The Iron Duke", which were machines built in the first half of the 19th century. Willshaw also mentions that there are many calenders, which are around a hundred years old, still providing good service.

Words and Expressions

crudely	[kru:dlɪ]	adv.	粗略的
precision	[prɪˈsɪʒn]	adj.	精确的,准确的
stack	[stæk]	v.	堆积
configuration	[kənˌfɪgəˈreɪʃn]	n.	布局,组态
tank lining			储(油)罐衬里
expansion joint			膨胀接头
imprecise	[ˌɪmprɪˈsaɪs]	adj.	不精确的
bubble	[ˈbʌbl]	n.	气泡
desirable	[dɪˈzaɪərəbl]	adj.	令人满意的
rough	[rʌf]	adj.	粗糙的
vertically	[ˈvɜːtɪklɪ]	adv.	垂直地;直立地
robust	[rəʊˈbʌst]	adj.	坚定的
solidly	[ˈsɒlɪdli]	adv.	坚固地;牢靠地
imposing	[ɪmˈpəʊzɪŋ]	adj.	壮观的,威风的
machinery	[məˈʃiːnəri]	n.	(总称)机器
monster	[ˈmɒnstə(r)]	n.	怪物
duke	[djuːk]	n.	公爵

Lesson 26 Compression Molding

Many rubber articles are produced by molding, a process in which uncured rubber—sometimes with an textile or metal insert—is cured under pressure in a mold. There are three general molding techniques: compression molding, transfer molding, and injection molding.

Compression molding is the simplest, cheapest, and probably the most widespread of the three basic molding techniques. It is ideally suited to small quantity production, say, from around fifty to a few thousand of each product annually. Figure 4.5 and Figure 4.6 show the various stages in the molding process. One of the keys to successful molding is adequate removal of air while the mold cavity is filling up with rubber.

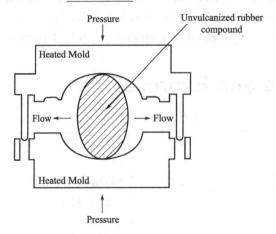

Figure 4.5 A loaded mold closing

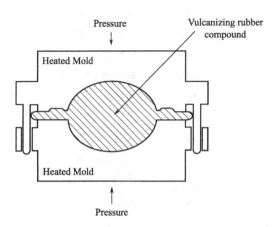

Figure 4.6 A loaded mold, closed under heat and pressure

The uncured pieces of compound placed in the mold are known-variously as preforms or load weights. For a ball, one might use an elliptically shaped extrusion, cut to an appropriate length. This shape is important and deliberately chosen so that air in the mold cavity will have a free path of escape when the mold begins to close, see Figure 4.5.

[1] Normally the weight of this preform will be chosen to be a few percent (2%~10%) above the weight of the final product, to ensure a fully formed product and to give an extra "push" for expulsion of any residual trapped air. The preform is placed in the bottom cavity and the top mold section placed on it by hand.

排除
残留的空气/模腔

The mold is continuously heated to a temperature, typically 120~180℃. A cure time for a smaller part might be 20 minutes, at 150℃, for thin cross-sections (less than 6 mm). In this case, temperatures above 150℃ could reduce the cure time to 10 minutes or less.

横截面

[2] The chemist plays his part in achieving a smooth flow of material in the mold, by striving to control the uncured compound viscosity. [3] This needs to be high enough to create the backpressure required to expel air efficiently as the mold closes, and low enough to permit completion of flow into all parts of the cavity before vulcanization begins. If we look at a low cured-hardness rubber, it usually contains little or no filler (NR & CR), or alternatively fillers plus a large quantity of oil. This can often make its viscosity too low for successful compression molding and the compounder may strive to increase its viscosity, by choosing a raw gum elastomer grade with a high Mooney viscosity. At the other end of the scale, high vulcanized-hardness compounds with lots of highly reinforcing fillers will need specialized process aids and low Mooney viscosity raw gum elastomers, to reduce viscosity, in order to promote the flow of the compound in the mold.

平稳流动

背压/有效地排除空气

低硫化硬度

生胶弹性体
门尼黏度
高硫化硬度胶料/补强填料/专用的/加工助剂
促进

As the press platens close the mold, excess compound begins to squeeze out into the grooves, taking air with it. [4] Often, residual air remains and various methods have been devised to remove it. One method is to bring the mold pressure back down to zero and then return to full pressure by quickly lowering and raising the press platens a number of times. This "shock" treatment is called "bumping". The shape of the preform and also its placement in the mold is important. The uncured rubber, placed in the cavity, might be a single piece or a number of pieces. This method is very much an art.

平板硫化机的压板
沟槽

模具压力

排气

The compression molding pressure is typically 7~10.5 MPa and will vary according to such things as the viscosity of the compound and the complexity of the mold cavity. The mold is designed to take the high stress involved.

胶料的黏度
模腔的复杂程度

It is interesting to note that the area of projected rubber can be smaller at the beginning of mold closure, since the rubber has not yet fully spread over all of the mold cavity. The flow of material in a mold is a complex process, especially in compression molding. [5] The rubber in the cavity is undergoing large temperature changes, which translate to viscosity variations thus continuously altering the flow characteristics of the compound. In recent years finite element analysis packages, which describe the material flow patterns in the mold, have become available to mold designers. The use of such design aids is at an early stage in most of the rubber industry.

流动特性
有限元分析
流动模型

New Words

compression	[kəm'preʃn]	n.	压缩,模压
transfer	[træns'fɜː(r)]	n.	转移
injection	[ɪn'dʒekʃn]	n.	注射
widespread	['waɪdspred]	adj.	普遍的;广泛应用
annually	['ænjuəli]	adv.	每年
adequate	['ædɪkwət]	adj.	足够的,充分的
removal	[rɪ'muːvl]	n.	除去
mold	[məʊld]	n.	模具
elliptically	[ɪ'lɪptɪkli]	adv.	椭圆地
deliberately	[dɪ'lɪbərətli]	adv.	深思熟虑地
escape	[ɪ'skeɪp]	n.	逃脱;泄漏
expulsion	[ɪk'spʌlʃn]	n.	排出;开除
residual	[rɪ'zɪdjuəl]	adj.	残余的;残留的
trapped	[træpt]	adj.	陷入困境的;受到限制的
cavity	['kævəti]	n.	腔,模腔
achieve	[ə'tʃiːv]	v.	取得;实现
expel	[ɪk'spel]	v.	排出(气体等)
completion	[kəm'pliːʃn]	n.	完成;实现
specialized	['speʃəlaɪzd]	adj.	专业的;专用的
promote	[prə'məʊt]	v.	促进,推进
squeeze	[skwiːz]	v.	挤,挤压
groove	[gruːv]	n.	沟,槽
devise	[dɪ'vaɪz]	v.	设计;发明
complexity	[kəm'pleksəti]	n.	复杂性
pattern	['pætn]	n.	模型,模式

Notes

[1] Normally the weight of this preform will be chosen to be a few percent (2%~10%) above the weight of the final product, to ensure a fully formed product and to give an extra "push" for expulsion of any residual trapped air. 通常情况下,选择的预制件的质量高于最终产品质量的百分之几(2%~10%),以确保产品的完全成型并给予额外的"推力"以排除残留的空气。

[2] The chemist plays his part in achieving a smooth flow of material in the mold, by stri-

ving to control the uncured compound viscosity. 化学家通过努力控制未硫化胶料的黏度来实现胶料在模具中的平稳流动。

[3] This needs to be high enough to create the backpressure required to expel air efficiently as the mold closes, and low enough to permit completion of flow into all parts of the cavity before vulcanization begins. （胶料在模具中的平稳流动）需要胶料具有适当高的黏度，以产生足够的背压，在模具关闭时能有效地排出空气；并且需要适当低的黏度以使得胶料在硫化开始之前能充满模腔。

[4] Often, residual air remains and various methods have been devised to remove it. One method is to bring the mold pressure back down to zero and then return to full pressure by quickly lowering and raising the press platens a number of times. 通常残留的空气仍然存在，并且已经设计出多种将其去除的方法。其中一种方法是将模具压力降回到零，然后通过快速降低和升高平板硫化机的压板并恢复到全压，反复数次（以排除空气）。

[5] The rubber in the cavity is undergoing large temperature changes, which translate to viscosity variations thus continuously altering the flow characteristics of the compound. 模腔中的橡胶经历巨大的温度变化，温度变化导致（胶料）黏度的变化，从而不断改变胶料的流动特性。

Exercises

1. Translate the following paragraphs into Chinese.

In compression molding, a pre-weighed, preformed piece is placed in the mold; the mold is closed, with the sample under pressure, as it vulcanizes. Cavity pressure is maintained by slightly overfilling the mold and holding it closed in the hydraulic press. Heat is provided by electric heaters, hot water or steam.

In the compression molding the weight, dimensions and positioning of the preformed compound have to be closely controlled, or the dimensions of the product can vary widely. This is often due to loss of material from the cavity as flash. This can be controlled, rather than be eliminated, by employing a shallow plunger which, due to close clearances, means that excess rubber can only escape when high pressure is applied, i.e., after the mold is completely filled.

2. Put the following words and phrases into English.

| 模压成型 | 传递模塑成型 | 注射成型 | 模具 | 模腔 |
| 硫化硬度 | 门尼黏度 | 加工助剂 | 流动特性 | 有限元分析 |

3. Please summarize the advantages of compression molding process.

[Reading Material]

Curing Equipment

Goodyear's discovery of vulcanization can be utilized. The rolls of sheeting have been calendered, the extrusions have been made, the Barwell has produced its preformed pieces, and shapes have been cut from milled sheet. The final step is to provide sufficient heat to change the uncured compound from a somewhat plastic state, to a dimensionally stable elastic substance, and additionally, in the case of molding, to achieve a final shape. An engineer would interpret curing as an increase in elastic modulus (G'). The chemist sees it as the formation of links between the chains, locking them together. The company owner sees it as a significant step in the transfer of money from a satisfied customer to the profit side of his financial ledger. Typical equipment used to achieve this change could be molds, autoclaves or even air curing ovens.

Molding

A mold might be described simplistically as at least two pieces of material (typically steel), which when fitted together form a cavity, resembling the shape of the product. This would be a very basic mold. Molding is by far the most important curing process, where uncross-linked rubber is placed into a heated mold, which gives it the final product shape, and then vulcanizes the material.

The mold

It can vary in size from a clenched fist to that of an automobile, and can have a single cavity to make one product at a time, or enough cavities to make a hundred or more. Most rubber molding is based on introducing a solid compound into a mold, although urethanes and silicones can be introduced as solids or liquids. It takes a fairly high mechanical pressure, to close the mold, and thus form the product shape; this pressure is provided by a press. Thus the mold must be strong enough to avoid being crushed. Tool steel hardened to a Rockwell C hardness of about 60 might be needed.

Mold design

A basic compression mold design is illustrated in Figure 4.7 which shows a cross-section through the center. Figure 4.8 shows the bottom half of a nine cavity mold. It is very important that the two halves of the mold register (fit accurately together). In this case, pins built into the top section fit snugly into holes drilled into the bottom half. Any looseness between the pin and the hole may cause the top half of the product to be out of alignment with the bottom half. If the fit is too tight, attempts to manually open the mold may prove difficult.

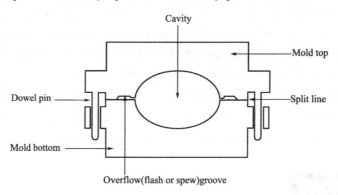

Figure 4.7　Closed empty compression mold

Since a number of compound materials expand with heat (the raw gum elastomer is of primary concern) by at least an order of magnitude more than steel, they will also shrink correspondingly as they cool when taken from the steel mold. Thus the mold dimensions are typically designed to be around 1.5% (based on linear dimensions) greater than those required in the rubber product, to compensate for the difference in expansion between the rubber and steel. This percentage vulcanizate shrinkage might be greater for FKM and silicone compounds and less for compounds with high amounts of filler.

Overflow grooves are machined around the mold cavity. In theory, this is to contain rubber

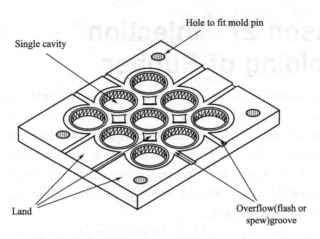

Figure 4.8　The bottom half of a nine cavity mold

in excess of the cavity volume. In practice for compression molds, it is not unusual to see during mold closure, material filling the cavity, then spilling out of the overflow grooves, and even across an area outside the grooves known as the land, and then out of the mold. This excess material is known as flash.

Introducing compound to the mold

There are different ways of introducing compound into the mold, some of which involve modifications to the basic design in Figure 4.7. They each confer certain advantages not found in the others. In the most basic design, (see Figure 4.7 and Figure 4.8, compression molding), pieces of rubber compound are placed in the bottom cavity and compressed using the top half of the mold.

Words and Expressions

autoclave	[ˈɔːtəkleɪv]	n.	高压釜
oven	[ˈʌvn]	n.	炉子；烘箱
air curing oven			空气硫化罐
resembling	[rɪˈzemblɪŋ]	v.	像…，类似于
crushed	[krʌʃt]	adj.	压碎的
pin	[pɪn]	n.	销钉
looseness	[ˈluːsnəs]	n.	松开；松度
alignment	[əˈlaɪnmənt]	n.	准线
groove	[gruːv]	n.	沟，槽

Lesson 27 Injection Molding of Rubber

Injection molding combines an extruder, which fluxes the rubber, with a reservoir and mold. While the rubber in the mold is curing, fresh rubber is prepared for the next cycle, so that molding is a continuous process. The compounds used have to be optimized for injection molding. They must flow through the nozzle and runner system and fill the mold under the pressure available. They should not cure before filling the mold, but cure quickly once in the mold.

Injection molding is now a well-established fabrication process in the rubber industry, its advantages over the older processes of compression and transfer molding can be summarized as:

(Ⅰ) Reduced labor costs;
(Ⅱ) Shorter cure times;
(Ⅲ) Better dimensional control of the product;
(Ⅳ) More consistent mechanical properties of the product.

[1] In specifying an injection molding system, it is necessary to decide on the type of injection machine, the size of the machine in terms of shot capacity, platen size, pressure, the number of mold cavities per platen, and the size of the gates and runners. The operation of an injection-molding machine requires the feeding, fluxing and injection of a compound, at a temperature close to the vulcanization temperature, in a measured volume, into a closed and heated mold, followed by a curing period, demolding, and, if necessary, mold cleaning, and/or metal insertion, before the cycle starts again. For maximum efficiency, as many of the above operations as possible should be automatic.

The Injection Molding process

In injection molding, rubber compounds are subjected to more severe processing conditions than that of compression molding. Temperatures, pressures, and shear stresses are higher, though cure times are shorter. Control over process variables can be more precise. [2] The cycle time can be minimized by independently controlling barrel temperature, screw speed, mold temperature, cure time, and injection pressure. Compounds with widely differing flow and cure characteristics can be molded into a variety of complex shapes and the skill lies in the optimization of the process.

The highest productivity is achieved when the compound is injected at a temperature close to the curing temperature, into a mold at a slightly higher temperature, because these conditions minimize both injection and cure time. [3] Scorch safety of the compound is the limiting factor in the process, and the effect of machine variables on compound temperature need to be understood if high injection temperatures and short injection times are to be achieved without scorch. Equally, it allows adjustments for batch-to-batch variations in viscosity, cure rate and scorch safety to be made if necessary.

Thus, to a large extent, controlling the injection molding process reduces to a question of homogeneity of temperature and heat history of the compound at each stage in the process. The temperature of the mix in the injection chamber, prior to injection, is determined by the temperatures of the extruder and injection chamber, by the screw speed, screw design, and by the applied back pressure. The injection, or mold filling, time and temperature depend on the temperature of the mix, as determined by the above factors, the injection pressure and the dimensions of nozzle, runners and gates. The cure time depends on the mold temperature and the temperature of the compound as it enters the mold.

Cold feed extruder designs have been optimized to enable control of the amount of frictional heat generated, and to maximize the overall heat-transfer rate by constantly exposing fresh rubber surfaces to both barrel and screw. Such turbulent flow also ensures thermal homogeneity of the entire volume of compound in the injection chamber, whose temperature is maintained as high as scorch safety permits.

Screw speed and design can both significantly affect heat generation in the extruder. In a given machine, screw design is fixed, and thus screw speed is the primary control factor. Plasticization time also depends on screw speed. A high screw speed can be used to minimize heat history by having the screw rotate only for the time required to fill the injection chamber.

Back pressure is controlled by the ram pressure, against which the screw must work whilst filling the injection chamber. Elevated back pressure raised the temperature of the mix and is only normally needed for low viscosity compounds which might not otherwise generate sufficient shear heat in their passage through the extruder.

[4] The work done by the ram in injecting the compound through the nozzle and runners is dissipated as heat, and this can boost the material temperature by over 40℃ above the injection chamber temperature. Thus, injection pressure as high as possible is used, which should consistent with freedom from scorch.

In summary, for safe injection molding the barrel and injection chamber temperatures, screw speed and back pressure are raised until the mix in the injection chamber is at the maximum safe temperature for the time it must remain there waiting to be injected. This time should be minimized by delaying the time of screw rotation. A nozzle diameter is then chosen that will permit a 5~10s injection time. Injection pressures and speeds as high as possible are then used consistent with freedom from scorch in nozzle, rubbers, gates and mold together with minimum mold filling times. Finally, mold temperatures can be raised to achieve minimum cure times. Thus it is clear, as stated earlier, that scorch safety of the compound is the limiting factor in injection molding.

New Words

flux	[flʌks]	v.	熔化;熔解
reservoir	['rezəvwɑː(r)]	n.	储槽
nozzle	['nɒzl]	n.	喷嘴
specify	['spesɪfaɪ]	n.	确定,规定
automatic	[ˌɔːtəˈmætɪk]	adj.	自动的
homogeneity	[ˌhɒmədʒəˈniːəti]	n.	均匀性
chamber	['tʃeɪmbə(r)]	n.	腔,室
frictional	['frɪkʃənəl]	adj.	摩擦的,摩擦力的
turbulent	['tɜːbjələnt]	adj.	湍流的
plasticization	[ˌplæstɪsaɪˈzeɪʃn]	n.	塑化
dissipate	['dɪsɪpeɪt]	v.	消散,消失
boost	[buːst]	v.	提高;增加

Notes

[1] In specifying an injection molding system, it is necessary to decide on the type of injection machine, the size of the machine in terms of shot capacity, platen size, pressure, the number of mold cavities per platen, and the size of the gates and runners. 在确定注射成型系统时，需要确定注射机的类型、注射机的尺寸、注射容量、压板尺寸、压力、每个压板的模腔数量以及注塑口型的尺寸和流道。

[2] The cycle time can be minimized by independently controlling barrel temperature, screw speed, mold temperature, cure time, and injection pressure. Compounds with widely differing flow and cure characteristics can be molded into a variety of complex shapes and the skill lies in the optimization of the process. 通过分别控制料筒温度、螺杆转速、模具温度、硫化时间和注射压力，可以将注射成型周期循环时间降至最短。流动性和硫化特性差别很大的胶料可以通过注射成型成为各种复杂形状的制品，其技巧在于优化注射工艺。

[3] Scorch safety of the compound is the limiting factor in the process, and the effect of machine variables on compound temperature need to be understood if high injection temperatures and short injection times are to be achieved without scorch. 胶料的焦烧安全性是该工艺中的限制因素，如果要在不发生焦烧的情况下实现高注射温度和短注射时间，则需要了解机器参数变化对胶料温度的影响。

[4] The work done by the ram in injecting the compound through the nozzle and runners is dissipated as heat, and this can boost the material temperature by over 40℃ above the injection chamber temperature. Thus, injection pressure as high as possible is used, which should consistent with freedom from scorch. 柱塞通过喷嘴和流道注入胶料消耗机械能将转化为热量消散，这可以将胶料温度提高至超过注射腔温度40℃以上。因此，使用尽可能高的注射压力，这应与保证焦烧自由性相一致。

Exercises

1. Translate the following paragraphs into Chinese.

In the standard injection process, the compound is injected into a closed mold. There are two variations of the process as described below.

(1) *Injection-compression molding*—In this process the mold is partially closed and a vacuum applied to the cavity area which is sealed by a compressible silicone gasket. A measured amount of rubber is injected into the partially opened mold. The mold is then closed and the excess rubber is forced outward to flow off channels. This process is used for cases where runner marks are unacceptable, e.g., precision O-rings.

(2) *Injection-transfer molding*—In this process the rubber is injected into a transfer chamber and then forced from the transfer chamber into the mold. This combination process uses the plasticization and heat generation advantages of the injection unit with the controlled flash pad and cavity layout advantages of the transfer press.

2. Put the following words and phrases into English.

注射成型	工艺条件	螺杆转速	模具温度	注射压力
硫化温度	料筒温度	焦烧安全性	限制因素	塑化时间
柱塞压力	喷嘴	流道	储槽	背压

3. What are the advantages of injection molding over the older processes of compression and transfer molding?

4. Translate the last paragraph of the text into Chinese.

[Reading Material]

The Advantages and Disadvantages of Injection Molding

An injection mold consists of a cylinder (injection barrel) with a ram or screw inside it, so that the rubber compound can be moved towards a nozzle at its end. The nozzle is then pressed against a hole made in the top half of a closed mold. This hole is then connected to smaller holes (gates and runners) which enter the cavities of the mold.

The compound can be presented to the barrel as a continuous strip, or in granulated form through a hopper, as in plastics injection molding. A ram has a tighter fit in the barrel than a screw and therefore there is less leakage backwards through the barrel; it is also cheaper than a screw. The screw "mixes" the compound as it moves towards the nozzle, creating more frictional heat and therefore higher temperatures which translate to easier flow and shorter cure times. A combination of ram and screw is popular.

Advantages

(a) The cure temperatures used can be much higher than those used for compression or transfer molding. For example, the injection temperature of NR is set as 160℃ and mold temperature of 180~190℃ as reported in some researchers' literature. High temperatures mean shorter cure times; one or two minutes are possible for thin cross-sections.

(b) Since the temperature of the compound entering the cavity is closer to the molding temperature, there is much less thermal volume expansion of the rubber during cure, therefore much less internal pressure build-up, resulting in a much reduced tendency to backrind.

(c) No complex preform is needed.

(d) Flash is significantly reduced or eliminated.

(e) Air entrapment is significantly reduced.

(f) The system is capable of a high level of automation, reminiscent of plastics injection molding.

(g) It is suited to fast, high quantity production.

Disadvantages

(a) The molds need to withstand very high injection pressures. This entails use of high hardness steel molds and higher precision tooling.

(b) If gates and runners are added to the mold its cost becomes significantly higher than compression or transfer molding.

Words and Expressions

cylinder	[ˈsɪlɪndə(r)]	*n.*	圆筒;气缸
injection barrel			注射料筒
granulated	[ˈɡrænjʊleɪtɪd]	*adj.*	颗粒状的
cavity	[ˈkævəti]	*n.*	腔,模腔
molding temperature			成型温度
thermal volume expansion			热容量膨胀
internal pressure build-up			内部压力聚集
entrapment	[ɪnˈtræpmənt]	*n.*	滞留

PART 5

Vulcanization of Rubber

Lesson 28 Definition

Vulcanization is the process by which rubber is changed from essentially a plastic material to either an elastic or a hard material. [1] As a plastic material unvulcanized rubber readily undergoes permanent deformation; whereas, vulcanized rubber is highly elastic and resistant to plastic deformation. Elastomeric rubber is much more commercially important than hard rubber (ebonite) and differs only in its lower content (0.5%~6%) of vulcanizing agent. (see Figure 5.1).

塑性材料
未硫化橡胶/永久变形

硫化剂

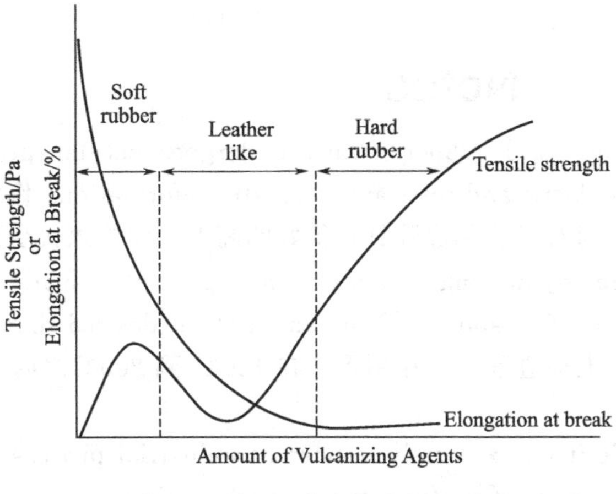

Figure 5.1 Influence of crosslinking on tensile strength and elongation at break of NR

[2] Rubber is normally vulcanized by heating with a vulcanizing agent (usually sulfur plus an accelerator) at a temperature between 120℃ and 200℃ in a mold of the desired shape and size. Vulcanized rubber is thermosetting since it cannot be remolded, in contrast to a thermoplastic material (e.g., polyethylene) which can be. The properties of rubber are improved dramatically by vulcanization. [3] This improvement has made it an extremely important industrial process, enabling the economical pro-

duction of a broad spectrum of useful articles, such as tires, shoe soles, electrical insulation, vibrational dampeners, etc. 电绝缘材料/减振器

　　Vulcanization involves the formation of chemical crosslinks, between the long polymer chains constituting the rubber molecules. As a result of the crosslink network the independent motion of the polymer chains is markedly restricted. Thus the chains cannot move past each other as readily as before vulcanization, with elongation decreasing and resistance to deformation (modulus) in-creasing. [4]Further, the crosslinks give the chains a much greater tendency to return to their original geometric arrangement when the deforming force is removed, which explains the greater elasticity and resistance to deformation. The three dimensional crosslinked structure renders vulcanized rubber more insoluble in solvents and more resistant to attack by heat, light and chemicals. 三维交联结构

New Words

ebonite	[ˈebənˌaɪt]	n.	硬橡胶
remold	[ˈriːˈməʊld]	v.	重塑
accelerator	[əkˈseləreɪtə(r)]	n.	促进剂
dimensional	[dɪˈmenʃənəl]	adj.	维的
render	[ˈrendə(r)]	v.	给予
geometric	[dʒiːəˈmetrɪk]	adj.	几何学的
insoluble	[ɪnˈsɒljəbl]	adj.	不溶的

Notes

　　[1] As a plastic material unvulcanized rubber readily undergoes permanent deformation; whereas, vulcanized rubber is highly elastic and resistant to plastic deformation. 作为一种塑料材料，未硫化橡胶容易产生永久变形；而对于硫化橡胶具有很好的高弹性和抗永久变形性。

　　[2] Rubber is normally vulcanized by heating with a vulcanizing agent (usually sulfur plus an accelerator) at a temperature between 120℃ and 200℃ in a mold of the desired shape and size. 橡胶硫化通常在硫化剂（一般用硫黄加上促进剂）的作用下，在120℃到200℃之间用加热的方式以获得所需的形状和尺寸。

　　[3] This improvement has made it an extremely important industrial process, enabling the economical production of a broad spectrum of useful articles, such as tires, shoe soles, electrical insulation, vibrational dampeners, etc. 这一改进使其成为一项极其重要的工业过程，能够经济地生产各种有用的物品，如轮胎、鞋底、电绝缘、减振器等。

　　[4] Further, the crosslinks give the chains a much greater tendency to return to their original geometric arrangement when the deforming force is removed, which explains the greater elasticity and resistance to deformation. 此外，当去除变形力时，交联使得分子链具有更大的恢复到原始几何形状的倾向，这解释了为什么硫化橡胶具有更大的弹性和抗变形性。

Exercises

　　1. Translate the first and second paragraphs of the text into Chinese.

2. Put the following words into English.

硫化橡胶 未硫化橡胶 塑性变形 弹性 促进剂
减振器 三维交联结构 不溶的

[Reading Material]

Sulfur in Rubber

The first and by far the most important cross linking agent is sulfur, which is relatively inexpensive and plentiful, and yet vital to the rubber industry. For a number of elastomers, the double bonds (unsaturation) discussed earlier are in plentiful supply on the polymer chain. Sulfur links one chain to another through these double bonds. Elastomers such as NR and SBR need only a small percentage of these double bonds to be utilized to produce a useful product; however this leaves the larger percentage unused and therefore vulnerable to attack by oxygen, ozone and heat.

Reaction of the small percentage of double bonds actually used for vulcanization can be achieved with 2~3 phr of sulfur in a conventional cure system. If more sulfur (30 phr) is added to the compound, many more of the available sites are cross-linked and the movement of the chains is so restricted after vulcanization that the rubber can barely stretch. It becomes very stiff (without the aid of any fillers) having a hardness of around 70 plus on the Shore D scale, and an inevitably reduced elongation at break of around 5%. The resulting material is called ebonite, so called because, when polished, it resembled the wood, ebony and indeed, before the days of plastic, polished ebonite was used for the handles of cutlery. Ebonite is still used today in tank linings and hair combs with a soft feel. The dense packing of cross links in ebonite reduces swelling of the rubber in liquids and the lower number of double bonds reduces attack by ozone and UV light. By far the best known ebonite is that produced from NR, although ebonites can easily be produced from both SBR and NBR.

Words and Expressions

vital to			对…关系重大
vulnerable	['vʌlnərəbəl]	adj.	易受攻击的
Shore D			邵氏 D 硬度
inevitably	[ɪn'evɪtəbli]	adv.	不可避免地
elongation	[ˌiːlɒŋ'geɪʃn]	n.	伸长率
resemble	[rɪ'zembl]	v.	类似于
ebony	[ebəni]	n.	乌木,黑檀
cutlery	['kʌtləri]	n.	刀具

Lesson 29 Measurement of Vulcanization

The changes which vulcanization brings to a sample of rubber are obvious, even to an untrained observer. The rubber loses its tack, becomes more elastic and has greater resistance to tensile, shear, compressive and torsional stress. The quantitative detection and measurement of vulcanization is much more difficult to perform, however.

[1] For many years rubber technologists followed the progress of the crosslinking reaction by curing different samples from a rubber mix for various times at a given temperature and then carrying out stress-strain measurements to find the tensile modulus (stress required to produce a given arbitrary elongation) and the tensile strength at break. [2] As the cure time was increased (and crosslinking progressed), the modulus and tensile strength increased until a plateau was reached at which they remained essentially constant as a function of cure time. The optimum cure time was taken as the minimum time with which these properties could attain the plateau. At best, this method is relatively crude. The error inherent in stress-strain measurements limits the quantitative accuracy and permits only gross changes in stock properties to be observed. The method cannot detect small differences between different stocks and different cures of the same stock.

The need for a better method of quantitatively measuring vulcanization has led to the development of a number of different curemeters. These devices provide a continuous measurement of the vulcanization process of a single sample of rubber. A curemeter measures the mechanical response of a stock to an oscillating stress as a function of time at cure temperature. A widely used curemeter is the oscillating disc type depicted schematically in Figure 5.2. In principle, the operation of other curemeters is quite similar to the oscillating disc type.

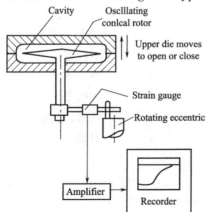

Figure 5.2 Schematic diagram of oscillating disc curemeter

The sample of rubber is placed in a closed cavity (see Figure 5.2) preheated to the cure temperature and containing a conical disc (rotor) which is continuously oscillated through a small arc of rotation. During cure the sample thus experiences a constant amplitude, oscillating torsional (or shear) stress. The <u>strain gauge</u> measures the resistance of the stock to the deformation imposed by the oscillating disc. This resistance, expressed as a torque, is then electronically amplified and transmitted to a recorder which plots torque (at the maximum oscillatory deformation) as a function of time. The resulting curve (see Figure 5.3) serves as a fingerprint for the specific sample of rubber being tested. Standard methods (ASTM D2084-95 and ISO 3417) are widely used for generating cure curves with the <u>oscillating disc rheometer</u>.

形变测量仪

旋转圆盘流变仪

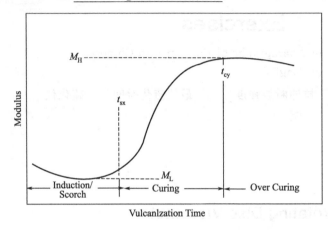

Figure 5.3 Cure curve from oscillating disc curemeter

The recent past has witnessed growing use of a <u>rotorless curemeter</u> in which the oscillations of the rotor have been replaced by those of the lower die. The cure curve from this rotorless curemeter is very similar to that from the older rotor instrument.

无转子硫化仪

New Words

torsional	['tɔːʃənəl]	adj.	扭转的
quantitative	['kwɒntɪtətɪv]	adj.	定量的
plateau	[plæ'təʊ]	n.	平坦时期
constant	['kɒnstənt]	adj.	持续的
oscillat	[ɒsə'leɪt]	v.	振荡
cavity	['kævəti]	n.	腔,模腔
preheat	[ˌpriː'hiːt]	v.	预热
conical	['kɒnɪkl]	adj.	圆锥(形)的
arc	[ɑːk]	n.	弧(度)
amplitude	[æmplɪtjuːd]	n.	振幅
rheometer	[rɪ'ɒmɪtə]	n.	流变仪

Notes

[1] For many years rubber technologists followed the progress of the crosslinking reaction by curing different samples from a rubber mix for various times at a given temperature and then carrying out stress-strain measurements to find the tensile modulus (stress required to produce a given arbitrary elongation) and the tensile strength at break. 多年来，橡胶技术人员通过在设定温度下对同一个样品进行不同时间的交联反应，然后对得到的试样进行应力-应变测试以确定其拉伸模量（产生单位伸长率所需的应力）和拉伸断裂强度。

[2] As the cure time was increased (and crosslinking progressed), the modulus and tensile strength increased until a plateau was reached at which they remained essentially constant as a function of cure time. 随着硫化时间增加（同时交联反应进行），模量和拉伸强度增加，直到达到硫化平坦期，在硫化平坦期这些性能随着固化时间基本上保持不变。

Exercises

1. Translate the first and second paragraphs of the text into Chinese.
2. Put the following words into English.

| 应力应变测试 | 拉伸模量 | 拉伸断裂强度 | 最佳硫化时间 | 硫化仪 |
| 硫化平坦期 | 弧（度） | 振幅 | | |

[Reading Material]

Rotating Disc Viscometer

An older, widely used instrument which complements the information from the curemeter is the rotating disc (Mooney) viscometer. The curemeter measures vulcanization properties; whereas, the Mooney viscometer measures processing and rheological properties.

The Mooney viscometer is based on a flat cylindrical die which rotates at a slow speed [typically 2 rpm (r/min)] in a closed cavity filled with the rubber sample to be tested (see Figure 5.4). The shearing action of the rotor meets a resistance from the viscous rubber which generates a thrust on the rotor shaft. This thrust is detected mechanically and transduced into an electronic signal which is plotted against time.

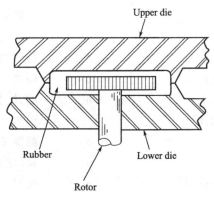

Figure 5.4 Rubber sample in Mooney viscometer cavity

Figure 5.5 gives a typical viscosity vs. time curve for a rubber stock in a Mooney viscometer.

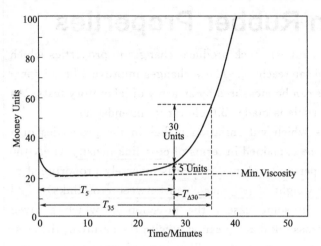

Figure 5.5 Mooney scorch curve

Words and Expressions

complement	[ˈkɒmplɪmənt]	v.	补充
rotating disc			旋转圆盘
viscometer	[vɪsˈkɒmɪtə]	n.	黏度计
Mooney viscometer			门尼黏度计
rheological	[riːəˈlɒdʒɪkəl]	adj.	流变学的

Lesson 30　Effects of Vulcanization on Rubber Properties

　　Vulcanization of a rubber stock produces changes in properties which are both profound and far-reaching. These changes improve a broad spectrum of properties, as can be measured by an array of laboratory tests. As a result, the rubber article is made suitable for its intended use.

　　[1] The changes which vulcanization makes in the properties of a rubber stock can best be explained in terms of crosslink density (i.e., the number of crosslinks per unit volume, which is inversely proportional to the average molecular weight of polymer chain between crosslinks). [2] The magnitude and, in some cases, the direction in which a given property varies with crosslink density can vary with the nature of the elastomer and the presence and amount of filler or reinforcing agent (e.g., carbon black, silica). However certain broad generalizations can be made.

交联密度

平均交联分子量

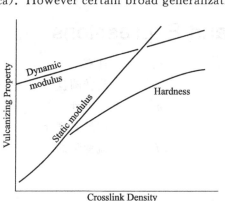

Figure 5.6　Variation of modulus and hardness with crosslink density in elastomeric region

　　It has been shown, both theoretically and experimentally that the modulus of a vulcanizate is proportional to the crosslink density. Thus as the number of crosslinks increases the elastomer chains are more restricted in their motion and the force required to produce a given deformation (modulus) becomes greater. Figure 5.6 depicts the approximately linear variation of modulus with crosslink density. Static modulus is that which would be measured by imposing a "slow" stress, as in an ordinary stress strain measurement. Dynamic modulus is that which is measured by the rapid application and removal of a deforming stress (usually in a sinusoidal manner). [3] The static modulus approximates the behavior of pure elasticity (i.e., conformance with Hooke's law, stress directly proportional to strain); whereas, the dynamic modulus reflects the viscoelastic properties (combination of viscous fluid and elastic material) of the rubber. It should be noticed that the static modulus passes

静态模量

动态模量

胡克定律

黏弹性能

through the origin whereas, the dynamic modulus is significant at zero crosslink density, reflecting the physical entanglements of the elastomer chains.

The increase in modulus with crosslinking is paralleled by a decrease in the elongation at break, a natural result of the progressively greater restriction of the polymer chains with increasing crosslinking. [4] The tensile strength on the other hand, rises to a maximum in the elastomeric (soft rubber) range, drops to a minimum in the leather like range and in the hard rubber range rises to values much higher than those of the soft rubber maximum. This unusual behavior is likely due to crystallization of the stretched natural rubber. With additional crosslinking, the restrictions on chain motion inhibit the crystallization and the tensile drops to a minimum. [5] Further crosslinking introduces additional chemical linkages between the individual chains, with the rubber behaving more and more like a solid and the tensile rising to values progressively higher than the maximum in the soft rubber stage. In the hard rubber stage the chains approach a saturation of crosslinks, with a highly elastic, low-deformation solid resulting.

Since hardness is resistance to indentation—a combination of compression, shear and tensile stress—It is not surprising that it progressively increases with crosslinking in approximately the same manner as modulus (see Figure 5.6). Both hardness and modulus are determined by the resistance to deformation.

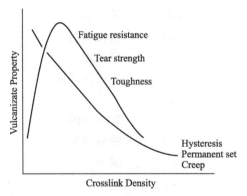

Figure 5.7 Variation of fatigue resistance, tear strength, toughness, hysteresis, permanent set and creep with crosslink density in elastomeric region

The elasticity of rubber is a measure of the reversibility of its deformation. A perfectly elastic material will return precisely to its original shape with no loss of mechanical energy as heat. Thus the hysteresis (ability to convert mechanical energy to heat) of rubber varies inversely with elasticity. [6] Formation of crosslinks progressively decreases the hysteresis shown in Figure 5.7, because the elastomer chain segments between crosslinks become increasingly shorter and the distance they can move on deformation becomes less. In many rubber applications high elasticity/low hysteresis is desirable (e.g., tires); in others (e.g., vibration damping) it is not. Changes in permanent set and creep parallel hysteresis.

Fatigue resistance, tear strength and toughness are determined by the energy to break, i.e., the area under the <u>stress strain curve</u>. At low levels of crosslinking these properties increase to a maximum (see Figure 5.7) and then decrease with additional crosslinks. 　　应力应变曲线

The influence of crosslink density on physical properties will vary with the nature of the elastomer and, to a lesser extent, with other compounding ingredients (e.g., reinforcing agents, cure system, plasticizer, etc.). It is possible, however, to make certain broad generalizations, crosslink density has a major influence on: Modulus (tensile, compression, shear, torsional), hardness, tensile strength, elongation at break, tear resistance, compression and tension set, creep, fatigue resistance, swelling resistance. The influence will be minor on: abrasion resistance, gas permeability, <u>glass transition temperature</u> (low 　　玻璃化转变温度
temperature flexibility), electrical conductivity, heat conductivity, chemical stability.

New Words

profound	[prəˈfaʊnd]	adj.	意义深远的
far-reaching	[fɑːˈriːtʃɪŋ]	adj.	深远的
inversely	[ˌɪnˈvɜːsli]	adv.	相反地
proportional	[prəˈpɔːʃənl]	adj.	成比例的
theoretically	[ˌθɪəˈretɪkli]	adv.	理论地
sinusoidal	[ˌsɪnəˈsɔɪdl]	adj.	正弦曲线的
conformance	[kənˈfɔːməns]	n.	一致
viscoelastic	[ˌvɪskəʊˈlæstɪk]	adj.	黏弹性的
hysteresis	[ˌhɪstəˈriːsɪs]	n.	滞后现象
creep	[kriːp]	n.	蠕变
indentation	[ˌɪndenˈteɪʃn]	n.	缺口
generalization	[dʒenrəlaɪˈzeɪʃn]	n.	一般化

Notes

[1] The changes which vulcanization makes in the properties of a rubber stock can best be explained in terms of crosslink density (i.e., the number of crosslinks per unit volume, which is inversely proportional to the average molecular weight of polymer chain between crosslinks). 根据交联密度（即每单位体积的交联点数量，其与平均交联分子量成反比），可以很好地解释硫化给橡胶性能带来的变化。

[2] The magnitude and, in some cases, the direction in which a given property varies with crosslink density can vary with the nature of the elastomer and the presence and amount of filler

or reinforcing agent (e.g., carbon black, silica). 在一些情况下，特定的性能随交联密度变化的大小和趋势是根据弹性体自身和补强填充剂（例如炭黑，二氧化硅）的性质相关的。

[3] The static modulus approximates the behavior of pure elasticity (i.e., conformance with Hooke's law, stress directly proportional to strain); whereas, the dynamic modulus reflects the viscoelastic properties (combination of viscous fluid and elastic material) of the rubber. 静态模量接近理想弹性的力学行为（即符合胡克定律，应力与应变成正比）；而动态模量反映了橡胶的黏弹性（既有弹性又有黏性的性质）。

[4] The tensile strength on the other hand, rises to a maximum in the elastomeric (soft rubber) range, drops to a minimum in the leather like range and in the hard rubber range rises to values much higher than those of the soft rubber maximum. 另一方面，（随着交联密度的增加）拉伸强度在高弹态（软橡胶）上升到最大值，随后在类似皮革的范围内下降到最小值，最后在硬质橡胶范围内上升到远高于高弹态的最大值。

[5] Further crosslinking introduces additional chemical linkages between the individual chains, with the rubber behaving more and more like a solid and the tensile rising to values progressively higher than the maximum in the soft rubber stage. 进一步的交联使大分子链之间产生了额外的化学交联键，这让橡胶越来越像固体，从而拉伸强度逐渐升高，最后超过高弹态橡胶的最大值。

[6] Formation of crosslinks progressively decreases the hysteresis shown in Figure 5.7, because the elastomer chain segments between crosslinks become increasingly shorter and the distance they can move on deformation becomes less. 如图 5.7 所示，交联的形成逐渐减弱了滞后现象，这是因为橡胶中交联点之间的分子链段变得越来越短，限制了链段在变形时的运动范围。

Exercises

1. Translate the first and second paragraphs of the text into Chinese.
2. Put the following words into English.

交联密度   平均交联分子量   黏弹性   动态模量   静态模量
滞后现象   蠕变   应力应变曲线

[Reading Material]

Crosslinking and Swelling

Crosslinking greatly reduces the effect of solvents on a specific rubber article. The greatest factor in this effect is the nature of the elastomer system and solvent. For a given elastomer in a given solvent the degree of dissolution of elastomer in the solvent and swelling of solvent into the elastomer will be determined by the degree of crosslinking. An uncrosslinked elastomer will completely dissolve in a good solvent. With crosslinking the solvent will only swell into the elastomer until the osmotic pressure of the solvent is equal to the elastic retractive forces of the stretched polymer chains. Any uncrosslinked portion of the elastomer will dissolve and diffuse out of the swollen rubber. The elastic retractive forces are inversely proportional to the weight of polymer chain between the points of crosslinking. Thus the swelling will decrease in proportion to the crosslink density. An elaborate theory has been developed by Flory to permit the calculation of crosslink density from the amount of a specific solvent imbibed by a vulcanizate.

Words and Expressions

solvent	[ˈsɒlvənt]	n.	溶剂
dissolution	[ˌdɪsəˈluːʃn]	n.	溶解
osmotic	[ɒzˈmɒtɪk]	adj.	渗透的
retractive	[rɪˈtræktɪv]	adj.	缩进的
portion	[ˈpɔːʃn]	n.	一部分
elaborate	[ɪˈlæbəret]	v.	详尽说明

Lesson 31 Vulcanization with Sulfur

Sulfur is still the most widely used curing agent. The use of sulfur was discovered independently by Goodyear (1839) in the United States and Hancock (1843) in Great Britain. Both observed that by heating natural rubber (NR) with sulfur its physical properties and aging resistance were markedly improved. A half century later (1893) Weber, a British chemist, discovered that the sulfur in vulcanized rubber was chemically combined rather than being simply dispersed. [1] Weber found that acetone extraction of NR previously heated with sulfur removed a fraction of the sulfur which progressively decreased with increasing heat treatment. This led to the conclusion that the extractible sulfur was simply dispersed in the NR, whereas the unextractible sulfur was chemically combined.

Modern concepts of polymer science permit a simple visualization of the chemical combination of sulfur with rubber. The long chains of the rubber molecules are simply linked together by crosslinks of one or more sulfur atoms. These crosslinks radically modify the properties of the rubber. Some of the sulfur chains connect different points of the same chain (intramolecular crosslinking), and others connect two points on different chains (intermolecular crosslinking). The change in properties resulting from vulcanization is due almost entirely to intramolecular crosslinking. [2] Intramolecular crosslinks have no effect, except in those instances where they aid the physical entanglement of adjacent chains.

The number of sulfur atoms per crosslink is of much importance. The efficiency of sulfur as a crosslinking agent varies inversely with the average number of S atoms (x) per crosslink. For a given amount of combined sulfur, the number of crosslinks will increase as x decreases. Of course, an actual vulcanizate will have a distribution of crosslinks with the various numbers of sulfur atoms ($x=1,2,3,4,\cdots,n$). [3] Chemical techniques have been developed for analyzing vulcanizates to determine the distribution of monosulfides ($x=1$), disulfides ($x=2$) and polysulfides ($x=3,4,5,\cdots,n$) in a given vulcanizate. The physical and chemical properties of a vulcanizate can be correlated with the relative amounts of mono-polysulfides, di-polysulfides and polysulfides.

Table 5.1 Dissociation energy of bonds in rubber crosslinks

Bond	Dissociation energy/(kcal/mol)
Alkyl—C—C—alkyl	80
Alkyl—C—S—C—alkyl	74
Alkyl—C—S—S—C—alkyl	54
Alkyl—C—S_n—S_m—C—alkyl	34

The variation of vulcanizate properties with the relative amounts of the various sulfides can best be understood in terms of the bond energy of the various bonds in the crosslinks. Table 5.1 gives a summary of the dissociation energies of the different types of chemical bonds found in sulfur crosslinks. These data lead to the conclusion that the C—C bond is more stable than the C—S, which, in turn is more stable than the S—S. Further, the stability of the S—S bond decreases as the number of sulfur atoms in the crosslink increases. Therefore, the chemical stability of sulfur crosslinks should decrease in the order.

  monosulfide＞disulfide ＞disulfide＞trisulfide＞tetrasulfide

  Rubber stocks with a high proportion of polysulfidic crosslinks are more susceptible to deterioration from heat and chemical attack. Consequently, on overcure they undergo much more reversion than those with high monosulfide content. Further, their retention of properties on aging is diminished. This observation is due to the lower stability of polysulfidic crosslinks. Both the tensile strength and elasticity of natural rubber have been found to be improved with polysulfidic content of crosslinks. They are poorest for C—C crosslinks ($x=0$). The improvement in tensile strength can be related to the greater flexibility and lability of the polysulfidic crosslinks. It appears that under tensile stress they can cleave and then reform. This process permits the relief of stress which would otherwise accumulate and lead to the initiation of failure.

  The lability of polysulfidic linkages also serves to explain the improvement in fatigue cracking resulting from their presence. In fatigue cracking tests the rubber sample is repeatedly strained to a specific strain below the ultimate limit. After a certain large number of cycles failure will occur due to crack formation and subsequent growth. In general, polysulfidic linkages will also favor resistance to abrasion, a property of basic importance in tire treads.

  Higher sulfur content in the crosslinks tends to increase the plasticity of the vulcanizate. Accordingly, the creep and compression set are greater for high polysulfidic stocks.

| 键能 |
| 化学稳定性 |
| 三硫键/四硫键 |
| 应力缓释 |
| 疲劳龟裂 |

New Words

extraction	[ɪk'strækʃn]	n.	萃取物
extractible	[ɪkst'ræktəbl]	adj.	可提取的
intramolecular	[ˌɪntrəmə'lekjʊlə]	adj.	分子内的
intermolecular	[ˌɪntə(ː)mə'lekjʊlə]	adj.	分子间的
dissociation	[dɪˌsəʊsɪ'eɪʃn]	n.	分离
susceptible	[sə'septəbl]	adj.	易受影响的
overcure	[əʊvə'kjʊə]	v.	过分硫化
reversion	[rɪ'vɜːʃn]	n.	恢复

| lability | [ləˈbɪlətɪ] | n. | 不稳定(性) |
| fatigue | [fəˈtiːg] | n. | 疲劳 |

Notes

[1] Weber found that acetone extraction of NR previously heated with sulfur removed a fraction of the sulfur which progressively decreased with increasing heat treatment. 韦伯用丙酮萃取加热后的含有硫黄的天然橡胶，发现萃取物中含有很少的硫黄，并且随着加热的程度加深得到的硫黄呈减少的趋势。

[2] Intramolecular crosslinks have no effect, except in those instances where they aid the physical entanglement of adjacent chains. 分子内交联没有效果，除了那些有助于增加相邻分子链的物理缠结的情况。

[3] Chemical techniques have been developed for analyzing vulcanizates to determine the distribution of monosulfides ($x=1$), disulfides ($x=2$) and polysulfides ($x=3, 4, 5\cdots$) in a given vulcanizate. 那些用于分析硫化橡胶以确定其中单硫化物（$x=1$）、二硫化物（$x=2$）和多硫化物（$x=3,4,5\cdots$）分布的化学技术已经被开发出来了。

Exercises

1. Translate the first and second paragraphs of the text into Chinese.
2. Put the following words into English.

| 单硫键 | 双硫键 | 多硫键 | 分子内交联 | 分子间交联 |
| 萃取物 | 应力缓释 | 疲劳开裂 | | |

[Reading Material]

Charles Goodyear

Charles Goodyear (December 29, 1800—July 1, 1860) was an American self-taught chemist and manufacturing engineer who developed vulcanized rubber, for which he received patent number 3633 from the United States Patent Office on June 15, 1844.

Goodyear is credited with inventing the chemical process to create and manufacture pliable, waterproof, moldable rubber. However, the Mesoamericans used a more primitive stabilized rubber for balls and other objects as early as 1600 BC.

Goodyear's discovery of the vulcanization process followed five years of searching for a more stable rubber and stumbling upon the effectiveness of heating after Thomas Hancock. His discovery initiated decades of successful rubber manufacturing in the Lower Naugatuck Valley in Connecticut, as rubber was adopted to multiple applications, including footwear and tires. The Goodyear Tire & Rubber Company is named after him.

Goodyear died on July 1, 1860, while traveling to see his dying daughter. After arriving in New York, he was informed that she had already died. He collapsed and was taken to the Fifth Avenue Hotel in New York City, where he died at the age of 59. He is buried in New Haven at Grove Street Cemetery.

Words and Expressions

self-taught	['self'tɔːt]	adj.	无师自通
patent	['pætnt]	n.	专利
pliable	['plaɪəbl]	adj.	柔韧的
Mesoamerican	[ˌmesəʊəˈmerikən]	n.	美索美洲
stumbling	['stʌmblɪŋ]	adj.	障碍的

Lesson 32 Non-Sulfur Vulcanization

Elastomers with chemically saturated polymer backbones cannot be crosslinked with sulfur and so require alternate curing agents. The most widely used of these are the <u>peroxides</u>, which can also be used with unsaturated elastomers. <u>Metal oxides</u> or <u>difunctional compounds</u> are used, as well, in special cases.

Peroxides cure by decomposing on heating into <u>oxy radicals</u> which abstract a hydrogen from the elastomer to generate a polymer radical. The polymer radicals then react to form carbon-carbon crosslinks. With unsaturated elastomers, this occurs preferentially at the site of <u>allylic hydrogens</u>. The rate of crosslinking is directly proportional to the rate of decomposition of the peroxide. Cure rates and curing temperatures therefore depend on the stability of the peroxide, which decreases in the order dialkyl>perketal>perester or diaryl. The most commonly used of these crosslinkers is <u>dicumyl peroxide</u>. Although carbon-carbon crosslinks are more thermally stable than sulfur crosslinks, they provide generally poorer tensile and tear strength. [1] Peroxides are also incompatible with many of the antioxidants used in rubber, since the antioxidants are designed to scavenge and destroy oxy and peroxy free radicals. Peroxides cannot be used with butyl rubber because they cause <u>chain scission</u> and depolymerization.

Certain difunctional compounds are used to crosslink elastomers by reacting to <u>bridge polymer chains</u>. [2] For example, diamines (e.g., <u>hexamethylenediamine carbamate</u>) are used as crosslinks for fluoroelastomers; <u>p-quinone dioxime</u> is oxidized to p-dinitrosobenzene as the active crosslink for bridging at the polymer double bonds of butyl rubber; and methylol terminated <u>phenol-formaldehyde resins</u> will likewise bridge butyl rubber chains (with $SnCl_2$ activation) as well as other <u>unsaturated elastomers</u>.

Metal oxides, usually zinc oxide but on occasion lead oxide for improved water resistance, are used as crosslinking agents for <u>halogenated elastomers</u> such as neoprene, halobutyl rubber, and chlorosulfonated polyethylene. The metal oxide abstracts the allylic halogen of adjacent polymer chains to form an oxygen crosslink plus the <u>metal chloride salt</u>.

过氧化物
金属氧化物/双官能度的化合物
含氧自由基

烯丙基氢

过氧化二异丙苯

断链

交联大分子链/六亚甲基二胺氨基甲酸酯/对醌二肟

酚醛树脂
不饱和弹性体

含卤弹性体

金属卤化物盐

New Words

decompose	[ˌdiːkəmˈpəʊz]	v.	分解
oxy	[ˈɒksɪ]	adj.	含氧的
preferentially	[ˌprefəˈrenʃəlɪ]	adv.	优先地
dialkyl	[daɪˈælkɪl]	n.	二烷基

perester	[pɜːˈrestɜː]	n.	过氧酯
diaryl	[daɪˈærɪl]	n.	二芳基
antioxidant	[ˌæntiˈɒksɪdənt]	n.	抗氧化剂
scavenge	[ˈskævɪndʒ]	v.	变坏，降解
scission	[ˈsɪʒən]	n.	分裂
diamine	[ˈdaɪəmiːn]	n.	二(元)胺
hexamethylenediamine	[heksæmeθɪləniːdaˈɪæmaɪn]	n.	己二胺
dioxime	[daˈɪɒksaɪm]	n.	二肟
methylol	[miːθɪˈlɒl]	n.	羟甲基

Notes

[1] Peroxides are also incompatible with many of the antioxidants used in rubber, since the antioxidants are designed to scavenge and destroy oxy and peroxy free radicals. 过氧化物与橡胶中使用的许多抗氧化剂是不能共用的，因为抗氧化剂旨在清除和破坏氧和过氧自由基。

[2] For example, diamines (e.g., hexamethylenediamine carbamate) are used as crosslinks for fluoroelastomers; p-quinone dioxime is oxidized to p-dinitrosobenzene as the active crosslink for bridging at the polymer double bonds of butyl rubber; and methylol terminated phenol-formaldehyde resins will likewise bridge butyl rubber chains (with $SnCl_2$ activation) as well as other unsaturated elastomers. 例如，二胺（例如，六亚甲基二胺氨基甲酸酯）用作氟橡胶的交联剂；对醌二肟氧化生成的对-二亚硝基苯与丁基橡胶分子中的双键有反应活性，可作为丁基橡胶的交联剂；用羟甲基封端的酚醛树脂可以交联丁基橡胶（在二氯化锡的催化下）和其他不饱和弹性体。

Exercises

1. Translate the first and second paragraphs of the text into Chinese.
2. Put the following words into English.

过氧化物　　金属氧化物　　双官能度的化合物　　抗氧化剂　　不饱和弹性体
含卤弹性体

[Reading Material]

Peroxide

Peroxide is a compound with the structure R—O—O—R. The O—O group in a peroxide is called the peroxide group or peroxo group. In contrast to oxide ions, the oxygen atoms in the peroxide ion have an oxidation state of -1. The most common peroxide is hydrogen peroxide (H_2O_2), colloquially known as "peroxide." It is marketed as a solution in water at various concentrations. Since hydrogen peroxide is colorless, so are these solutions. It is mainly used as an oxidant and bleaching agent. Concentrated solutions are potentially dangerous when in contact with organic compounds.

Peroxide curing is widely used for curing silicone rubber. The curing process leaves behind byproducts, which can be an issue in food contact and medical applications. However, these products are usually treated in a postcure oven which greatly reduces the peroxide breakdown product content. One of the two main peroxides used, dicumyl peroxide, has principal breakdown products of acetophenone and phenyl-2-propanol. The other is dichlorobenzoyl peroxide, whose principal breakdown products are dichlorobenzoic acid and dichlorobenzene.

Words and Expressions

peroxo group			过氧基团
oxide ion			氧离子
oxidation state			氧化态
colloquially	[kə'ləʊkwɪəlɪ]	adv.	用通俗语
bleaching agent			漂白剂
byproduct	['baɪˌprɒdʌkt]	n.	副产品
postcure	[pəʊstkjʊə(r)]	n.	后固化
acetophenone	[æsɪtə'fenəʊn]	n.	苯乙酮
dichlorobenzene	[daɪklɔːrəʊ'benziːn]	n.	二氯(代)苯

Lesson 33 Dynamic Vulcanization

In recent years, a great deal of attention has been focused on recycling of both commodity and engineering polymers. Different methods, such as chemical recycling, reclaiming and reutilization have been developed for the disposal of polymer waste. [1]Large-scale recycling of plastic and rubber waste could be done by utilizing them in thermoplastic vulcanizates (TPVs) produced from dynamically vulcanized rubber/plastic blends. Properties of these materials depend upon the concentration of the regrind, as well as on the adhesion between the different polymeric phases. 热塑性硫化橡胶

Dynamic vulcanization is the selective crosslinking of rubber and its fine dispersion in a molten thermoplastic via an intensive mixing and kneading process. This process yields a fine dispersion of partially or fully crosslinked micrometer size rubber particles in a thermoplastic matrix (see Figure 5.8). These dynamically vulcanized rubber/plastic blends are known popularly as TPVs, and are processable like thermoplastics. They can also be recycled without much deterioration in properties. Dynamic vulcanization improves elasticity and stabilizes the morphology of TPVs. [2] Typical property improvements achieved by dynamic vulcanization are: reduced permanent set, improved ultimate mechanical properties, improved fatigue resistance and resistance to aggressive media. [3] TPVs provide high value-added products if the components are derived from waste sources, and thus utilization of waste plastics and/or rubber in TPVs is known as "upcycling". 动态硫化

永久变形
抗疲劳性能/腐蚀性介质

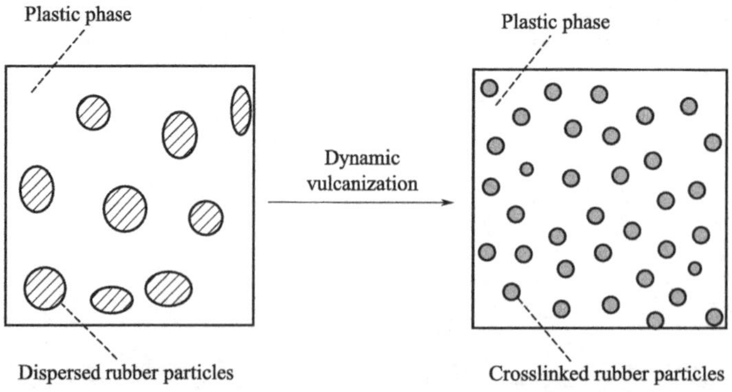

Figure 5.8 Schematic representation of the morphology of a TPV

New Words

reclaim	[rɪˈkleɪm]	v.	回收再利用
reutilization	[riːjuːtɪlaɪˈzeɪʃən]	n.	回收利用

regrind	[riː'graɪnd]	v.	将…再研磨
knead	[niːd]	v.	揉捏
micrometer	[maɪ'krɒmɪtə(r)]	n.	千分尺
processable	['prəʊsesəbl]	adj.	可加工的

Notes

[1] Large-scale recycling of plastic and rubber waste could be done by utilizing them in thermoplastic vulcanizates (TPVs) produced from dynamically vulcanized rubber/plastic blends. 通过动态硫化的方法用它们制备橡胶/塑料共混的热塑性硫化橡胶（TPVs），可以实现废旧塑料和橡胶的大规模回收。

[2] Typical property improvements achieved by dynamic vulcanization are: reduced permanent set, improved ultimate mechanical properties, improved fatigue resistance and resistance to aggressive media. 动态硫化的方法可以减轻永久变形性能显著提高材料的力学性能、抗疲劳性能和抗腐蚀性介质的性能。

[3] TPVs provide high value-added products if the components are derived from waste sources, and thus utilization of waste plastics and/or rubber in TPVs is known as "upcycling". 如果主要原料来自废旧资源，热塑性硫化橡胶可制备高附加值产品，因此 TPVs 中废塑料和/或橡胶的利用被称为"升级版再利用"。

Exercises

1. Translate the first and second paragraphs of the text into Chinese.
2. Put the following words into English.

动态硫化　　　抗疲劳性能　　抗永久变形性　　废旧塑料　　废旧橡胶
热塑性硫化橡胶　　回收

[Reading Material]

TPEs and TPVs

The production of multicomponent materials such as blends and composites manufactured from the most inexpensive large-tonnage polyolefins (polyethylene, polypropylene, and their copolymers) still have an important role in the modern materials science. Thus, owing to the unique combination of mechanical and rheological behaviors of the components in a final product, thermoplastic elastomers (TPEs), based on PP and various elastomers, are among the most widely used industrial materials. TPE combines the advantageous mechanical properties of conventional rubbers with an easy processability inherently belonging to linear thermoplastic polymers above their melting temperatures.

The properties of TPEs can be substantially improved using the method of dynamic vulcanization. In this technique, the elastomer vulcanization and its mixing with thermoplastic component occur simultaneously at elevated temperature (i.e., beyond the melting of the thermoplastic component). TPEs prepared by dynamic vulcanization are identified as thermoplastic vulcanizates

(TPVs). TPV is a material in which crosslinked rubber particles are present in a relative high amount 50%～70% (wt) and finely dispersed form (1～3μm in diameter) in the continuous thermoplastic matrix. TPVs based on PP and ethylene-propylene-diene monomer (EPDM) are the most important materials from the commercial and scientific points of view.

Words and Expressions

multicomponent	[mʌltɪkəm'pəʊnənt]	adj.	多组分的
large-tonnage	['lɑːdʒt'ʌnɪdʒ]	adj.	大产量
rheological behavior			流变性能
processability	[prəʊsesə'bɪlɪtɪ]	n.	加工性能
simultaneously	[ˌsɪməl'teɪnɪəslɪ]	adv.	同时地

PART 6
Property Tests of Uncured and Cured Rubber

Lesson 34 Introduction of the Basic Concepts of Testing

Everyone understands the need for testing. The effort and expense of testing are borne in order to create a level of confidence that some function will be fulfilled as intended/needed. Some kinds of testing are comparatively simple; all the concepts involved are easy to understand and the results are easy to interpret.

There are four concepts that are fundamental to any kind of testing. These are validity, accuracy, precision, and reproducibility.　　　　　　　　　　有效性/准确度/精密度/可重复性

(1) Validity—[1]A test is valid if whatever is being done will yield information about the performance characteristic of interest.

(2) Accuracy—Once the correct characteristic is being measured, the next question is how good the measurement is.

(3) Precision—[2]When a test is run more than once on supposedly identical or at least very comparable specimens, in theory the results will　　试样
always be the same. In practice this only happens when:

A. the specimens are truly identical (more or less a statistical impos-　　完全相同
sibility unless it is the exact same specimen in a nondestructive test);　　无损检测

B. the test method is not sensitive enough to detect the small random variations in properties that must exist even in very comparable specimens;

C. the test method itself is not subject to any scatter in its measurements (also statistically unlikely).

The variations from specimen to specimen and from one test occasion to the next will always combine to produce some scatter in the output　　测试结果
of the test. (This is formally known as testing error, but it does not mean　　测试误差
anybody made a mistake.) [3] This means that when a test is run repeatedly there will be a certain range in the results even when the test specimens are thought to be reasonably uniform. The spread of that range

demonstrates how precise the test method is for that kind of specimen. Really good tests have narrow ranges of results, perhaps ±2%~3%, other tests may run more like ±50% or more.

Test precision is very important if small sample sizes are to be used. In a worst case, if a single sample of each of two different rubber compounds were <u>tensile tested</u>, and one had a <u>tensile strength</u> of 2000 psi and the other tested at 1800 psi, someone might conclude that the first compound was in fact the stronger. But tensile strength of rubber is subjected to at least ±10% scatter, so for a single point comparison as described, the apparent difference of 200 psi in strength could not be relied upon to really be meaningful. It would be perfectly possible for a retest of a sample group of 5~10 specimens of each compound to show their average strengths were essentially the same. 拉伸测试/拉伸强度

[4] Probably the single most common pitfall of testing is lack of knowledge about the precision of the particular test method being used, which leads to <u>judgments</u> that some batch of material or product is different from some other batch when it is really not. 判断

Reproducibility—Just as there can be <u>variations</u> in running the same test from day to day in a laboratory, there can be and often are variations in performing the same test procedure in different laboratories. 变化

[5] Reproducibility in a test method means that it can be run in many laboratories and all the results will be comparable. Very often this is not a problem, but sometimes a particular combination of <u>test equipment</u> and details of the method is such that there are substantial disagreements when attempts are made to duplicate test procedures in multiple laboratories. 测试设备

[6] Concerns over reproducibility and general understanding of testing variation are sometimes addressed by having a group of samples made as identically as possible and then broken into numerous subgroups which are dispersed to different participating laboratories to be subjected to the same test procedure in each. This is often referred to as <u>round robin testing</u>, and the data generated can be very useful in comparing laboratories. 循环测试

Therefore, ideally a test method will be valid, accurate, precise, and reproducible. It would be extremely convenient if the method was also easy, quick, and inexpensive to run, but generally if it can be described by any one of those terms the technologist would be content. However, not all test procedures share those four characteristics, and not every laboratory is run with full scientific rigor all the time, so the value and meaning of test data can and should be questioned at times.

Many of the test methods published by the American Society for Testing and Materials (ASTM) have included in them studies on their precision and reproducibility, which are usually the least read sections of the method. Examining those sections is strongly recommended.

New Words

concept	[ˈkɒnsept]	n.	概念，观念
interpret	[ɪnˈtɜːprɪt]	v.	理解；解释
validity	[vəˈlɪdəti]	n.	有效性
accuracy	[ˈækjərəsi]	n.	准确（性）
precision	[prɪˈsɪʒn]	n.	精确度
reproducibility	[rɪprədjuːsəˈbɪlɪti]	n.	可重复性
identical	[aɪˈdentɪkl]	adj.	完全同样的，相同的
comparable	[ˈkɒmpərəbl]	adj.	可比较的
specimen	[ˈspesɪmən]	n.	样品；试样
statistical	[stəˈtɪstɪkl]	adj.	统计的；统计学的
conclude	[kənˈkluːd]	v.	得出结论；推断出
judgment	[ˈdʒʌdʒmənt]	n.	判断
variation	[ˌveəriˈeɪʃn]	n.	变化
duplicate	[ˈdjuːplɪkeɪt]	v.	重复；复制
rigor	[ˈrɪɡə]	n.	严格；严酷

Notes

[1] A test is valid if whatever is being done will yield information about the performance characteristic of interest. 如果正在进行的测试能得出所需的性能特征的信息，则测试是有效的。

[2] When a test is run more than once on supposedly identical or at least very comparable specimens, in theory the results will always be the same. 当使用相同或至少非常相似的试样进行（同一个实验的）多次测试时，理论上测试结果应该总是相同的。

[3] This means that when a test is run repeatedly there will be a certain range in the results even when the test specimens are thought to be reasonably uniform. The spread of that range demonstrates how precise the test method is for that kind of specimen. 这表明即使测试试样相当均匀，当重复进行测试时，结果也会有一定的范围区间。该范围的扩展证明了试样的该种测试方法的精确度。

[4] Probably the single most common pitfall of testing is lack of knowledge about the precision of the particular test method being used, which leads to judgments that some batch of material or product is different from some other batch when it is really not. 最常见的测试问题（困难）可能是缺乏对所使用的特定测试方法的精确性的知识，这导致错误地判断某批材料或产品与其他批次（材料或产品）不同，而它们确实没有什么不同。

[5] Reproducibility in a test method means that it can be run in many laboratories and all the results will be comparable. Very often this is not a problem, but sometimes a particular combination of test equipment and details of the method is such that there are substantial disagreements when attempts are made to duplicate test procedures in multiple laboratories. 测试方法的可重复性表示它可以在许多实验室中进行，且所有结果都具有可比性。通常这不是问题，但有时测试设备和方法细节的特定组合使得在多个不同实验室中尝试重复的测试程序时测试结果存在一定的

差别。

[6] Concerns over reproducibility and general understanding of testing variation are sometimes addressed by having a group of samples made as identically as possible and then broken into numerous subgroups which are dispersed to different participating laboratories to be subjected to the same test procedure in each. 通过将一组样品制成尽可能相同的样品，然后将其分成许多小组，分散到不同的参与测试的实验室进行相同的测试程序，从而解决对重复性和测试差异的担忧。

Exercises

1. Translate the last two paragraphs of the text into Chinese.
2. Put the following words and phrases into English.

| 有效性 | 准确性 | 精确度 | 可重复性 | 试样（样品） |
| 相同的 | 可比较的 | 测试误差 | 测试设备 | 循环测试 |

[Reading Material]

Introduction of Rubber Compounds Testing

Most testing of rubber and rubber compounds is conducted to determine processing characteristics or to measure physical properties after vulcanization. Processability of a rubber compound is dependent on the compound's viscosity and elasticity. Generally, the physical properties of vulcanized rubber compounds are measured by static and dynamic mechanical tests designed to simulate the mechanical conditions of finished rubber products.

The fabrication of rubber products generally involve the mixing and processing of unvulcanized compounds through complex equipment. Tests to measure the processability of unvulcanized rubber are chiefly concerned with rheological properties. That is, the response of the rubber compound to the forces and temperatures imposed on it during the operations of mixing, extrusion, calendering, and curing. Generally, these tests are used for control purposes in factory operations to ensure that subsequent processing and curing steps are carried out uniformly. The earliest and still one of the most popular rheological instruments is the shearing disk viscometer.

Words and Expressions

processability	[prəʊsesə'bɪlɪtɪ]	n.	加工性能
dynamic mechanical test			动态力学性能测试
rheological	[riːə'lɒdʒɪkəl]	adj.	流变学的
viscometer	[vɪs'kɒmɪtə]	n.	黏度计
shearing disk viscometer			剪切盘式黏度计

Lesson 35 Selected Compound Properties and Test

[1] Laboratory tests on rubber compounds are run to determine properties, and, to some extent, to predict service life of products. The purpose here is to briefly mention type of tests needed for certain specific properties. 　　使用寿命

Tests on uncured compounds—Compound viscosity is measured by using the Mooney viscometer and gives an indication of how a compound might process. 　　未硫化胶料 门尼黏度计

Curemeters are used to determine compound scorch time for processing safety and cure rate at a specified temperature. 　　硫化仪/焦烧时间 加工安全性/硫化速率

Tests on cured compounds—The most commonly measured property of rubber is tensile strength which is the ultimate strength of rubber. The elongation at which the sample breaks is the ultimate or breaking elongation. Stress at a specific elongation is known modulus at that elongation. Typically rubber industry looks at moduli at 100% and 300% elongation although most rubber products seldom experience more than 20%~25% elongation in service. It is suggested that rubber technologists and engineers look at low strain modulus for design purposes. 　　断裂伸长率 应力   低应变模量

The most common aging tests include hot air aging, ozone aging and then measuring residual tensile strength and elongation. 　　热空气老化/臭氧老化

Compounds used under dynamic conditions need to be tested for flex resistance. Fuel hose compounds should be tested for fluid (fuel) resistance. Tire tread compounds are additionally tested for abrasion, tear resistance and traction. Common rubber tests and their ASTM standards are listed in Table 6.1. 　　耐屈挠性 耐介质性 抗撕裂性能 牵引力

Table 6.1 Rubber compound tests and ASTM designations

Type	Compound property	Test	ASTM designation
Uncured	Viscosity	Mooney viscometer	D1646-96
	Scorch time and cure time	Rheometer	D2084-95
Cured	Tensile str., elong., modulus	Tensile tester	D412-87
	Hardness	Durometer hardness	D2240-97
	Flex resistance	DeMattia flex tester	D430-95
	Compression resistance	Compression set	D395-87(94)
	Abrasion resistance	Pico abrasion	D228-88(94)
	Tear resistance	Die C tear test	D624-91
	Fluid resistance	Fluid immersion test	D471-95
	Air aging resistance	Hot air aging test	D573-88(94)
	Ozone resistance	Static ozone test	D518-86(91)
		Dynamic ozone test	D3395-86(94)

New Words

elongation	[ˌiːlɒŋˈgeɪʃn]	n.	伸长
stress	[stres]	n.	应力
modulus	[ˈmɒdjʊləs]	n.	模量
flex	[fleks]	n.	弯曲,屈挠
abrasion	[əˈbreɪʒn]	n.	磨损
traction	[ˈtrækʃn]	n.	牵引力

Notes

[1] Laboratory tests on rubber compounds are run to determine properties, and, to some extent, to predict service life of products. The purpose here is to briefly mention type of tests needed for certain specific properties. 在实验室对胶料进行测试以确定其性能，并在一定程度上预测产品的使用寿命。这里的目的是简要介绍某些特定性能所需的测试类型。

Exercises

1. Translate the second, third and fourth paragraphs of the text into Chinese.
2. Put the following words and phrases into English.

使用寿命	门尼黏度计	硫化仪	焦烧时间	硫化速率
加工安全性	热空气老化	臭氧老化	耐屈挠性	耐介质性

[Reading Material]

Introduction of Flex Resistance

The term flex resistance does not mean resistance to flexing or resistance to bending. It means the ability to withstand numerous flexing cycles without damage or deterioration. Flex-crack resistance means the ability to sustain numerous flexing cycles without the occurrence of cracks in the surface resulting from stress and ozone attack. Flex-cut-growth resistance means the ability to withstand numerous flexing cycles with a cut in the stressed surface with little or no growth of that cut.

The most common type of failure or damage resulting from repeated flexing is the formation of surface cracks, known as flex cracking. Shoe soles and tires are examples of rubber products that may show flex cracking. Another mode of failure due to flexing is the separation of rubber from fabric in a fabric-supported product such as a conveyor belt.

The two most commonly used flex resistance tests are the Dematia Flex Test (ASTM D813) and the Ross Flex Test (ASTM D1052). The Dematia flexer alternately pushes the test specimen together, bending it in the middle, then pulls it back to straighten it. The Ross flexer repeatedly bends the specimen over a metal rod, then straightens it. In both tests, a small cut or nick of prescribed size and shape may be made in the center of the test specimen. Data generated from these tests include: flex life, the number of test cycles a specimen can withstand before it reaches a specified state of failure; and crack growth rate, the rate at which the cut propagates itself as the sample is flexed. These tests are particularly useful in evaluating compositions that are intended for use in products which will undergo repeated flexing or bending in service.

Another, more elaborate, test is the Monsanto Fatigue to Failure Tester. This instrument measures the ultimate fatigue life as cycles to failure. Tensile-like samples are flexed or stretched at 100 cycles per minute at a preselected extension ratio. Samples can be extended over a range of 10% to 120%. The fatigue performance of compounds can be measured and compared either at constant extension ratio (strain) or at constant strain energies (work input). This instrument is more likely to be used as a research and development tool than as a factory control tester.

Words and Expressions

flex	[fleks]	n.	弯曲,屈挠
flex resistance			耐屈挠性
deterioration	[dɪˌtɪərɪəˈreɪʃn]	n.	恶化;变坏
flex-crack resistance			耐屈挠龟裂性
surface crack			表面龟裂
flex cracking			屈挠龟裂
conveyor belt			传送带
flex life			屈挠寿命
crack growth rate			裂口增长速度
Monsanto Fatigue to Failure Tester			孟山都疲劳失效测试仪
constant extension ratio			恒定伸长率

Lesson 36　Viscosity Tests

The first tests in the manufacturing process for rubber products are those which evaluate the raw elastomer to determine its quality and assess its capacity to be successfully compounded. Foremost among these tests are the viscosity (or plasticity) tests.

The tests used to measure viscosity can be classified in terms of shear rate. This is a logical basis for dividing the discussion since the various rubber processes—mixing, molding, calendering, extrusion—all involve different shear rate levels. This is graphically illustrated in Figure 6.1. Figure 6.1 shows the influence of shear rate on shear stress for two compounds. Although the two compounds rank in one order at very low shear rates, the order is reversed at higher shear rates. Thus the two compounds may have the same flow characteristics at the lower shear rate encountered in compression molding, but markedly different flow behavior at the higher shear rates encountered in extrusion. [1] As a result, tests to measure the processing characteristics of a compound should expose the sample to shear rates similar to those encountered in the actual process. This is not always possible, since the shear rate achieved by many test instruments is lower that the shear rate of the process being simulated. [2]To compensate, test measurements are often made at temperature lower than the process temperature since for most compounds lowering the test temperature is approximately equivalent to increasing the shear rate.

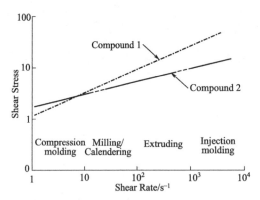

Figure 6.1　Effect of shear rate on shear stress.

A number of the rotary shear plastometers were developed during the period of rapid growth of the rubber industry in 1920—1940. However, the shearing disk viscometer developed by Melvin Mooney[11] has become the "standard" rotary shear instrument of the rubber industry. Mooney viscosity specifications are now contained in most rubber material purchasing agreements. [3] In this instrument, a flat, serrated disk rotates in a massed rubber specimen contained in a grooved, heated cavity under pressure, as shown in Figure 6.2. The torque required to rotate the disk at 2rpm (r/min) at a fixed temperature (usually 100℃) is defined as the Mooney viscosity.

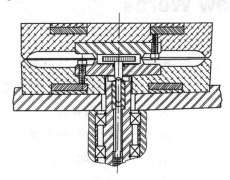

Figure 6.2 Platens, dies and rotor in typical shearing disk viscometer

The Mooney viscometer operates at an <u>average shear rate</u> of about $2s^{-1}$. This is <u>an order of magnitude</u> higher than compression plastometers, but still within the range of only compression molding operations. 平均剪切速率
一个数量级

The test procedure involves placing a sample on either side of the rotor and filling the cavity by pneumatically lowering the top platen. The platens are electrically heated and controlled at a preset temperature. The sample is allowed to warm up for 1 minute after the platens are closed. The motor is then started and a <u>torque transducer</u> measures the torque required to turn the rotor at 2 rpm. Test data includes <u>initial torque</u> and torque after fixed time periods as specified in the test procedure being executed or in ASTM Standard D1646. 转矩传感器
初始转矩

The Mooney viscometer is strictly an empirical instrument measuring torque in arbitrary "Mooney" units. Another imprecise feature of this instrument is the measurement temperature. Since the rotor is unheated, the true measurement temperature is only implied by the platen temperature. Per D1646, Mooney results from a typical test are usually reported as follows:

$$50-ML\ 1+4(100℃)$$

Where 50 refers to viscosity in Mooney units, M indicates Mooney, L indicates use of the <u>large "Mooney" rotor</u>, 1 is the time in minutes that the specimen was preheated, 4 is the time (minutes) after starting the motor at which the reading is taken, and 100℃ is the set temperature of the platens. 大门尼转子

There are two standard rotors for the Mooney viscosity test: a large rotor for general use and a small rotor for very stiff materials. The ratio of the viscosity results with the two rotors (L/S) is about 1.8, but it varies with the type rubber being tested.

The Mooney viscometer is also used to measure the scorch characteristics and stress relaxation of fully compounded rubber.

焦烧特性
应力松弛

New Words

viscosity	[vɪˈskɒsətɪ]	n.	黏度
plasticity	[plæˈstɪsətɪ]	n.	可塑度(性);塑性
logical	[ˈlɒdʒɪkl]	adj.	逻辑(上)的;符合逻辑的
compensate	[ˈkɒmpenseɪt]	v.	抵消;补偿
plastometer	[plæsˈtɒmɪtə]	n.	塑性仪,塑度仪
rotary	[ˈrəʊtərɪ]	adj.	旋转的
standard	[ˈstændəd]	adj.	标准的
specification	[ˌspesɪfɪˈkeɪʃn]	n.	规格;说明书
disk	[dɪsk]	n.	圆盘
rotor	[ˈrəʊtə(r)]	n.	转子
torque	[tɔːk]	n.	转矩;(机器的)扭转力
magnitude	[ˈmægnɪtjuːd]	n.	量级
pneumatically	[njuːˈmætɪkəl]	adv.	由空气作用
empirical	[ɪmˈpɪrɪkl]	adj.	凭经验的,经验主义的

Notes

[1] As a result, tests to measure the processing characteristics of a compound should expose the sample to shear rates similar to those encountered in the actual process. 因此,在胶料加工特性的测试中,应使样品处于与实际工艺中承受的剪切速率相似的剪切速率。

[2] To compensate, test measurements are often made at temperature lower than the process temperature since for most compounds lowering the test temperature is approximately equivalent to increasing the shear rate. 为了抵消(剪切速率的差异),测试通常在低于加工温度的温度下进行,因为对于大多数胶料来说,降低测试温度相当于增加剪切速率。

[3] In this instrument, a flat, serrated disk rotates in a massed rubber specimen contained in a grooved, heated cavity under pressure, as shown in Figure 6.2. 在这台仪器中,一个扁平的边缘呈锯齿状的圆盘(转子)携带一定质量的橡胶样品,在一定压力下于凹槽形的加热腔中旋转,如图6.2所示。

Exercises

1. Translate the sixth and seventh paragraphs of the text into Chinese.
2. Put the following words and phrases into English.

黏度	可塑度	黏度测试	剪切速率	加工特性
转子	塑性仪	应力松弛	焦烧特性	橡胶制品

3. Please explain the meanings of the parameters in "50-ML 1+4(100℃)".

[Reading Material]

Scorch Testing

Scorch, previously defined as premature vulcanization, is really a vulcanization property. However, since it is extremely important in defining processability limits, it will be briefly considered here. It is obvious that the viscosity of a fully compounded stock held at elevated temperature will increase (albeit nonlinearly) with time as a result of crosslinking. Thus, continuous measurement of viscosity at processing temperatures will indicate the time available for further processing. In this regard, the Mooney viscosity test described in ASTM D1646 is almost universally used to determine the scorch characteristics of compounded rubber. The time required for a compound to scorch is determined from an analysis of the Mooney units vs. time plot shown in Figure 6.3. The most common measure of scorch is the time to a 5-point rise above the minimum as shown in Figure 6.3. Normally the test is run at temperatures encountered during processing, i.e., between 120℃ and 135 ℃. Values derived from this Mooney cure curve are defined below:

MV = Minimum viscosity
t_5 = Time to scorch   at $MV+5$ units
t_{35} = Time to cure   at $MV+35$ units
Δt_L = cure index $= t_{35} - t_5$

Bear in mind that these characteristics define bulk scorch properties since the compound is treated as a uniform mass.

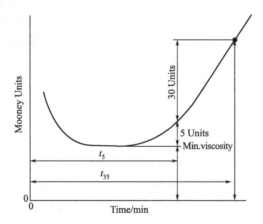

Figure 6.3 Typical Mooney Viscometer curing characteristics curve

Scorch, or premature vulcanization, may occur during the processing of a compound due to accumulated effects of heat and time. Therefore the time required for the compound to scorch will decrease as it moves through each stage in the process. Thus, samples taken from the same batch at different stages in the process will have progressively shorter scorch times. The value for a batch at a particular point in the process is called its "residual scorch time". Curemeter curves on "before processing" and "after processing" samples are shown in Figure 6.4 to illustrate this heat history effect. An acceptable factory stock must have a scorch time slightly longer than the time equivalent of the maximum heat history it may accumulate during processing. It is important to emphasize that the curves presented in Figure 6.4 were obtained with a curemeter rather than a Mooney instrument as the curemeter allows one to look at the entire cure curve.

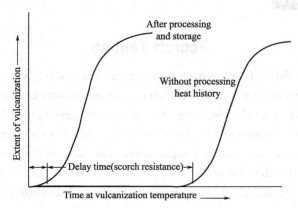

Figure 6.4　Effect of heat history on scorch safety

Words and Expressions

processability	[prəusesə'bılıtı]	n.	加工性能
scorch characteristic			焦烧特性
Mooney cure curve			门尼硫化曲线
minimum viscosity			最低黏度
cure index			硫化指数
scorch safety			焦烧安全性
scorch time			焦烧时间
residual scorch time			剩余焦烧时间

Lesson 37 Vulcanization Testing

If the processability testing indicates that the stock has not scorched, it is now suitable for conversion to a final product. [1] This step involves shaping the product then applying additional heat for a period of time to vulcanize the polymer backbone and fix its shape. This latter process transforms the material from one that is viscoelastic to one that retains its final shape. [2] It was previously pointed out that vulcanization involves chemically crosslinking the polymer chains into an elastic network that can be deformed by an applied stress but will return to its initial shape when this stress is removed. The progress of this vulcanization reaction can be assessed by measurement of the retracting force which opposes a mechanical deformation. On this basis, temperature controlled tests have been developed to assess the compounds performance throughout the vulcanization process and to get an early assessment of the product's final properties.

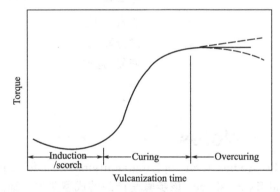

Figure 6.5 Rheometer cure curve showing stages in the vulcanization

The physical changes during the process occur in three stages: (1) the induction or precure period; (2) the curing or crosslinking stage; and (3) an overcure or reversion stage. These stages are illustrated in Figure 6.5. During the induction period chemical reactions occur which increase the viscosity of the compound but don't result in significant crosslinking or network development. Thus the compound still behaves as a fluid. The curing phase, following the induction period, is then crosslinking proceeds rapidly. As the curatives and crosslink sites are depleted, the crosslinking reactions slow until an optimum stiffness or modulus is achieved. Throughout this stage the compound acts as an elastic solid. Further heating may not change the stiffness or may result in a slow increase in stiffness ("marching modulus") as shown by the upward curving in Figure 6.5. Alternatively continued heating may result in stock softening ("reversion") as shown by the downward curving in Figure 6.5. Heating beyond the optimal stiffness is referred to as overcure. The relative time involved in each of these phases is of course a function of temperature, rubber type, and formulation.

A rubber technologist attempts to balance the three stages in the design of a rubber compound/process. [3] The formulation must allow sufficient scorch time to get the compound through the <u>forming stage</u> before the material changes from a shapeable fluid to an elastic solid; yet the curing reaction should be accelerated as much as possible to minimize the mold time required to obtain <u>optimal cure</u>. This balance is achieved through rubber compounding and assessed via vulcanization tests.

成型阶段

最佳硫化

New Words

viscoelastic	[ˌvɪskəʊˈlæstɪk]	adj.	黏弹性的
retain	[rɪˈteɪn]	v.	保持
assess	[əˈses]	v.	评定;确定
assessment	[əˈsesmənt]	n.	评估;确定
rheometer	[rɪˈɒmɪtə]	n.	流变仪
induction	[ɪnˈdʌkʃn]	n.	诱发,诱导
overcure	[ˌəʊvəˈkjʊə]	n.	过分硫化
reversion	[rɪˈvɜːʃn]	n.	恢复,返原
phase	[feɪz]	n.	相
proceed	[prəˈsiːd]	v.	进行

Notes

[1] This step involves shaping the product then applying additional heat for a period of time to vulcanize the polymer backbone and fix its shape. 该步骤是产品的成型阶段，通过对胶料加热一段时间，将聚合物（橡胶）主链硫化并固定其形状。

[2] It was previously pointed out that vulcanization involves chemically crosslinking the polymer chains into an elastic network that can be deformed by an applied stress but will return to its initial shape when this stress is removed. 之前已经指出，硫化是将聚合物链通过化学交联形成弹性网络，该弹性网络能在应力作用下发生形变，但是当该应力去除时能回复到其初始形状。

[3] The formulation must allow sufficient scorch time to get the compound through the forming stage before the material changes from a shapeable fluid to an elastic solid; yet the curing reaction should be accelerated as much as possible to minimize the mold time required to obtain optimal cure. 所设计的配方需有足够的焦烧时间，在胶料从塑性流体转变为弹性固体之前保证胶料能够成型；应具有尽可能快的硫化反应速度，以保证（胶料）在获得最佳硫化的条件下硫化时间最短。

Exercises

1. Translate the second paragraph of the text into Chinese.
2. Put the following words and phrases into English.

| 硫化 | 硫化测试 | 硫化反应 | 黏弹性 | 弹性网络 |
| 预硫化期 | 硫化返原期 | 过硫化 | 流变仪 | 硫化相 |

3. Please describe the definition of vulcanization in English.

[Reading Material]

Testing Procedure of Curing Characteristics

The oscillating disc curemeter or rheometer (ODR) solves the problem of not being able to make any rheological measurements after the scorch time (as with the Mooney viscometer), by changing the rotor from a rotating mode to an oscillating one. Since cured rubber can stretch to some extent without breaking, the oscillations are kept within this limit. The magnitude of the oscillation is measured in degrees of arc, 1° and 3° are most common, and the rate of oscillation is suggested as 1.7Hz. The curemeter is an essential piece of equipment and used extensively in the rubber laboratory. The machine plots a graph of torque verses time for any given curing temperature. The full extent of cure and beyond can now be recorded. For example reversion, the point at which the vulcanized compound breaks down due to prolonged heating can now be measured.

During the testing, a piece of uncured compound rubber is placed on the heated rotor, and the heated top die cavity is immediately brought down on to the lower die thus filling the cavity. In Figure 6.6, the curve shows an immediate initial rise in torque upon closure of the heated cavity. At the top of this first "hump", the compound has not had much chance to absorb heat from its surroundings, and since viscosity is temperature dependent it will be somewhat higher in these first few seconds. As the compound absorbs heat from the instrument, it softens. Its temperature then stabilizes, and its viscosity has a constant value prior to the onset of cure. This assumes that it is not masked by a very short scorch time. This is the first important feature on the curve. It is the minimum viscosity of the rubber at the chosen temperature and degree of oscillation; and it has the symbol M_L.

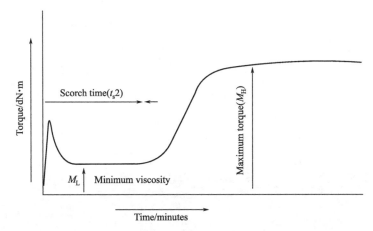

Figure 6.6 Typical oscillating disc curemeter curve

After a certain time, the viscosity (torque) begins to increase, indicating that the curing process (vulcanization or cross linking) has begun. The time from the closure of the cavity to this moment is the next important property, the scorch time. It has the symbol t_s1, which means, the number of minutes to a 1dN·m rise above M_L, (used with a 1°arc). If a 3° arc is chosen, then a scorch time with the symbol t_s2 is used, which is a 2dN·m rise above M_L.

The torque continues to increase, until there is no more significant rise. At this point the compound is vulcanized, and this maximum torque value is designated by the symbol M_H.

The last major piece of information to be extracted is the time it takes to complete the cure, known as the cure time. The symbol for this property is t'_x. This is defined with some precision as "the time taken for the curve to reach a height expressed as the value of M_L plus a "percentage" of the difference between M_H and M''_L. If we think of this "percentage" as a decimal, then 90% is 0.9 (this number is commonly chosen), and mathematically this height would be expressed as $0.9(M_H - M_L) + M_L$. This "cut off" at 90% is often known as the technical cure time of the compound, and has the symbol t'_{90}.

Words and Expressions

oscillate	[ɒsɪleɪt]	v.	振荡
oscillation	[ˌɒsɪˈleɪʃn]	n.	振荡
oscillating disc curemeter			振荡盘式硫化仪
rheological measurement			流变(特性)测量
essential	[ɪˈsenʃl]	adj.	基本的；必要的
constant value			恒定值
decimal	[ˈdesɪml]	n.	小数
technical cure time			工艺正硫化时间

Lesson 38 Tensile Test

Tensile testing of rubber is preferably performed on standard specimens, which are dumbbell shapes die cut from a standard molded test sheet, six inches square by about 0.080 inches (1in＝0.0254m). thick, as shown in Figure 6.7. However, sometimes the test is done using other specimens, such as O-rings; or a nonstandard specimen might be made from a sheet of material which has been somehow cut or ground from an actual rubber article.

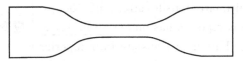

Figure 6.7 Dumbbell shaped test piece for tensile testing

The test specimen is secured in the jaws of a tensile test machine and stretched at a specified rate until it breaks. The final force required for the break is recorded, along with the amount of stretch that was achieved at the break point. The forces in effect at various degrees of elongation of the specimen are also usually recorded. These forces are used to calculate the stresses per unit area at those elongations, which are reported as tensile modulus. These are usually abbreviated, so that the stress at 100% strain is referred to as the 100% tensile modulus, or the M-100. Stresses at higher strains are then referred to as the M-200, M-300, etc.

[1] It is very important to understand that none of these numbers is a classic modulus, that is, a basic ratio of stress to strain that applies across a wide range of strains. [2]With the possible exception of a narrow region of moderately low strain, the stress-strain plot for elastomers is always nonlinear, and these are secant moduli drawn to various points of the particular curve that applies at that temperature and rate of strain. In Figure 6.8 a typical stress-strain plot is displayed, with a line drawn on it to illustrate what a secant modulus is.

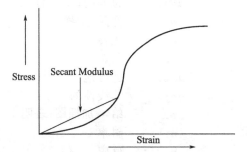

Figure 6.8 Stress-strain plot for a rubber specimen

Engineering calculations sometimes require use of a quantity called Young's Modulus. For rubber this may be determined from the slope of the stress-strain line at very low strains (5%).

[3] For a few materials, such as steel, a comparatively pure modulus can be measured, which is not a function of strain, temperature, or rate of strain. It is a single point of information, so to speak, that applies very broadly. In contrast, the response of rubber to a deforming force is not a point, it is a three dimensional surface, the axes of which are temperature, amount of strain (deformation), and rate of strain.

If two compounds are tested at the same time and one ruptures at 500% elongation, with a final stress of 3000psi (20.68MPa)—measured against the original cross-sectional area of the test specimen—while the other fails at 300% elongation and ultimate tensile stress of 1800psi (12.4MPa), the judgment that the first compound has higher strain tolerance and greater strength will be valid. This may translate into superior fitness for use for some application, but not necessarily, especially if the two compounds are based on different polymers.

The tensile moduli mentioned above will have more meaning in comparisons of compounds. It is safe to say that if a compound has its average M-100 (100% modulus) even 10%~15% higher than that of another compound, the first rubber is significantly stiffer than the other. For many applications there will be a relationship between the tensile modulus of the compound and the stiffness of the final rubber part, although it will seldom be a very precise relationship.

So the usefulness of tensile test data ranges from moderately applicable (M-100 in relation to part stiffness), to questionable in meaning (elongation in relation to flex fatigue), to indirectly useful (very low tensile strength as an indicator of poor compound quality). [4] However, tensile data have been used historically as an important part of quality control and the test is almost invariably called out in any specification covering rubber materials.

函数
变形力
断裂
原始的横截面积
应变耐受性
拉伸模量
屈挠疲劳
质量控制

New Words

tensile	['tensaɪl]	adj.	拉伸的；张力的
specimen	['spesɪmən]	n.	样品，抽样
dumbbell	['dʌmbel]	n.	哑铃
secured	[sɪ'kjuəd]	n.	固定
moderately	['mɒdərətli]	adv.	适度地
nonlinear	['nɒn'lɪnɪəl]	adj.	非线性的
illustrate	['ɪləstreɪt]	v.	给…加插图；表明
slope	[sləʊp]	n.	斜率
response	[rɪ'spɒns]	n.	反应
rupture	['rʌptʃə(r)]	n.	断裂，破裂
judgment	['dʒʌdʒmənt]	n.	判断
superior	[suː'pɪərɪə(r)]	adj.	（在质量等方面）较好的

moduli	[ˈmɒdʒəˌlaɪ]	n.	模量（modulus 的复数形式）
questionable	[ˈkwestʃənəbl]	adj.	可疑的
specification	[ˌspesɪfɪˈkeɪʃn]	n.	规范；说明书；标准

Notes

[1] It is very important to understand that none of these numbers is a classic modulus, that is, a basic ratio of stress to strain that applies across a wide range of strains. 理解这些数值（M-100，M-200）中没有一个是典型的模量是非常重要的，也就是说，应力与应变的基本比例适用于较广范围的应变。

[2] With the possible exception of a narrow region of moderately low strain, the stress-strain plot for elastomers is always nonlinear, and these are secant moduli drawn to various points of the particular curve that applies at that temperature and rate of strain. 除了低应变的较狭窄区域外，弹性体的应力-应变曲线始终是非线性的；那些绘出的特定曲线的各个点的割线模量，适用于该温度和速率下的应变。

[3] For a few materials, such as steel, a comparatively pure modulus can be measured, which is not a function of strain, temperature, or rate of strain. 对于少数材料，例如钢，可以测量相对较纯粹的模量，该模量不是应变、温度或应变速率的函数。

[4] However, tensile data have been used historically as an important part of quality control and the test is almost invariably called out in any specification covering rubber materials. 然而，拉伸数据历来被用作质量控制的重要组成部分，并且在橡胶材料的任何规范中几乎总是要求测试（其拉伸性能）。

Exercises

1. Translate the following four paragraphs into Chinese.

Stress-strain curves are generated through the pulling (or tensile straining) of a "dumbbell" sample (ASTM D412). This dumbbell is died out of a flat cured sheet about 2mm thick. The stress required to break the dumbbell is the tensile strength. The strain or % stretch registered at the instant of rupture is called ultimate elongation, or simply elongation. It is common practice to record and tabulate the stress values at 100%, 200%, etc., instead of plotting curves.

Tensile properties are normally measured at room temperature (about 23℃). Equipment is available for doing this at lower or higher temperatures, which have a pronounced effect on the hardness and stiffness of a vulcanizate.

Tensile stress, more commonly called Modulus is the stress required to produce a certain strain, or elongation. Modulus values can be taken at any elongation off the stress/strain curve. In rubber testing, modulus is usually reported as 100% modulus, 300% modulus, or 600% modulus, meaning the stress at 100%, 300%, or 600% elongation.

Tensile strength is an important characteristic as a tool for compound development, manufacturing control and determination of susceptibility to deterioration by oil, heat, weather, and other environmental factors. It is not necessarily an indication of quality, but is often used as such. It is true that the low range of tensile strength values is indicative of low quality compounds and the high range indicative of high quality. However, in the mid-range (about 1500~3000psi, 1psi=6894.76Pa), it is difficult to correlate tensile strength with other mechanical properties.

2. Put the following words and phrases into English.

拉伸强度	拉伸测试	拉伸模量	100%定伸应力	杨氏模量
应力-应变曲线	标准试样	特定速率	横截面积	质量控制

[Reading Material]

Significance of Tensile Testing

An important function of tensile testing is to determine how well ingredients are dispersed in the rubber compound, during the mixing stage. For example, if carbon black is poorly dispersed, the tensile strength (at break) of the cured compound will be lower than it should be. A low state of cure, due to insufficient curative, as well as inadequate cure time or temperature, will also give a lowered tensile strength.

If a compound has too much carbon black, not enough oil, or too high a state of cure, perhaps due to excessive sulfur or accelerator, it will be reflected in a higher modulus value. A severely over processed NR compound might have a lowered modulus value. Compounds used in the rubber industry have tensile strengths from less than 7MPa to around 28MPa. Urethanes can have even higher tensile strengths. There are cases where tensile strength is specifically relevant to an application, for example an elastic band. A higher tensile strength is also preferred for highly dynamic applications.

Words and Expressions

mixing stage			混炼阶段
urethanes	[ˈjʊərɪθeɪnz]	n.	聚氨酯橡胶
specifically	[spəˈsɪfɪkli]	adv.	特有地,特别地
elastic band			橡皮筋

Lesson 39 Durometer Hardness

[1] A basic test of surface hardness is attractive in its simplicity and relative ease of performance, given the availability of rugged, pocketsize devices. For rubber what is measured is resistance to penetration by a small indentor which is spring loaded. When the durometer is carefully pressed against the rubber surface, a relative measure of this resistance is observed on the durometer face, using a scale of 0~100, although the accuracy and precision of the measurements become questionable towards either end of the scale. As with other materials, no single device will adequately measure this property across its full possible range, which for rubber includes soft foams and extremely hard urethanes. Therefore several scales exist, ranging from the Shore OO for foams to the Shore D for the very hard rubbers. The most commonly applied scale is the Shore A, which will read a soft baby bottle nipple around 30 and a hard shoe sole around 80. There is also the International Rubber Hardness Durometer (IRHD) scale, which is very similar to the Shore A.

[2] There is a rough correlation between stiffness of a rubber compound and its surface hardness. If the same type of polymer is used for compounds, the correlation improves; if the compounds are all basically the same, only varying in ingredients that affect their stiffness (such as carbon black and process oil), the correlation becomes fairly good. However, assumptions are often made that are inappropriate, such as equating compounds in stiffness if their durometers are the same or very close.

[3] Since two compounds with the same durometer can vary easily by 25% or more in actual stiffness, and/or be affected in stiffness very differently as temperature or other factors vary, such an assumption can badly mislead the technologist or engineer.

However, hardness will remain a basic and easily understood way of describing rubber materials, which does convey significant information about them.

New Words

durometer	[djuəˈrɒmɪtə]	n.	硬度计
hardness	[hɑːdnəs]	n.	硬度
simplicity	[sɪmˈplɪsəti]	n.	简单
penetration	[ˌpenɪˈtreɪʃn]	n.	渗透；穿透
indentor	[ɪnˈdentə]	n.	硬度计的压头
nipple	[ˈnɪpl]	n.	橡胶奶嘴
sole	[səʊl]	n.	鞋底

inappropriate	[ˌɪnəˈprəʊprɪət]	adj.	不恰当的，不适宜的
equating	[ɪˈkweɪtɪŋ]	v.	认为某事物（与另一事物）相等或相仿
assumption	[əˈsʌmpʃn]	n.	假定，假设

Notes

[1] A basic test of surface hardness is attractive in its simplicity and relative ease of performance, given the availability of rugged, pocketsize devices. 对于鉴别坚固耐用的便携式设备的可用性，表面硬度的基本测试在其简单性和相对容易的性能方面具有吸引力。

[2] There is a rough correlation between stiffness of a rubber compound and its surface hardness. If the same type of polymer is used for compounds, the correlation improves; if the compounds are all basically the same, only varying in ingredients that affect their stiffness (such as carbon black and process oil), the correlation becomes fairly good. 橡胶胶料的硬度与其表面硬度之间存在粗略的相关性。如果胶料中使用相同高聚物（生胶），那么相关性会提高；如果胶料基本相同，只是影响其硬度的配合剂种类不同（如炭黑和加工油），那么相关性会变得相当好。

[3] Since two compounds with the same durometer can vary easily by 25% or more in actual stiffness, and/or be affected in stiffness very differently as temperature or other factors vary, such an assumption can badly mislead the technologist or engineer. 由于两种具有相同硬度值的胶料的实际硬度可轻易相差25%或更多，并且/或者当温度或其他因素变化时，对硬度产生非常不同的影响，像这种假设可能会严重误导技术专家或工程师。

Exercises

1. Translate the following paragraph into Chinese.

By far the most frequently measured rubber properties are hardness, tensile strength, and elongation. These are often referred to as the physical properties of a vulcanizate. These are supplemented by additional tests designed to be more or less predictive of actual service performance and service life.

2. Put the following words and phrases into English.

| 硬度 | 硬度计 | 硬度计压头 | 表面硬度 | 国际橡胶硬度 |
| 邵氏 A | 邵氏 A 硬度 | 橡胶奶嘴 | 鞋底 | 相关性 |

[Reading Material]

Hardness

Hardness, as applied to rubber products, is the relative resistance of the surface to indentation under specified conditions. Hardness of rubber is usually measured with a small spring-loaded hardness gauge known as a durometer (ASTM D2240). The durometer may be handheld or mounted on a stand. The measurement is made by pressing the indentor against the sample and reading the scale, which is calibrated in arbitrary units ranging from 0 (soft) to 100 (hard). A Type A durometer is used for most soft rubber products; there is also a Type D durometer for hard rubber and plastic-like materials. On the A scale, a gum rubber band would measure around 40, a tire tread around 60, and a shoe sole around 80.

Hardness is probably the most often-measured property of elastomer vulcanizates. It appears in almost every specification and is widely used by rubber goods manufacturers for quality control. However, its practical significance is questionable, because surface indentation rarely bears any relationship to end use performance. Also, hardness is a very imprecise measurement; it is not uncommon to find 5 or more points difference in readings by different people on the same piece of rubber. Nevertheless, hardness will probably remain as a standard test in the rubber industry because it is simple and quick to measure and has been correlated with product quality over many years of experience.

Words and Expressions

indentation	[ˌɪnden'teɪʃn]	n.	压痕
gauge	[geɪdʒ]	n.	测量仪器
practical significance			实际意义
end-use performance			最终使用性能
imprecise	[ˌɪmprɪ'saɪs]	adj.	不精确的

Lesson 40　Compression Set

　　In numerous applications a rubber part is subjected to either a constant load (ASTM D395 Method A, very seldom used) or a constant deformation (Method B). The rubber motor mount which supports a car's engine is a good example of the first situation, and an O-ring seal is the classic example of the second. [1] In both cases it would be desirable for the rubber to continue to fulfill its function in exactly the same way over time; that way the position of the engine would not change in relation to the car chassis, and the O-ring would exert the same sealing force after months or years of service as it did when first installed.

　　[2] But elastomers are viscoelastic, which means they will store energy reversibly (elasticity) as a coiled spring does, but also that they will dissipate energy through viscous behavior, like a shock absorber. Viscous behavior means that molecules undergo some form or another of flow or rearrangement. This amounts to stress relaxation, a time dependent phenomenon. [3] In practice, this guarantees that overtime the motor mount will deflect a little more (drift or creep), allowing the engine to move closer to the ground, and the O-ring will exert less sealing force than originally. In either case, if the rubber part is freed from its load/deformation, it will not entirely regain its original geometry.

　　The most common method of compression set testing uses a molded button shape, a half-inch in height and one square inch in cross-section. The button sample is either subjected to a constant load from a strong spring (Method A) or compressed 25% in a metal fixture (Method B). After the test condition has been maintained for a time and temperature appropriate to the rubber and the application, the button sample is removed from stress and allowed to recover. [4] The difference between the recovered dimensions of the rubber and its original dimensions is referred to as set, and the amount of set is an indicator of how much stress relaxation the rubber underwent while subjected to the load or the deformation. [5] In the case of the constant imposed deformation, if the rubber did not recover at all but retained the dimensions imposed during the test, that would be a set of 100%, which would of course correlate with complete stress relaxation.

　　However, there is not necessarily a tight correlation between the amount of set measured and the level of stress relaxation. Two O-rings might both display a 50% set after a Method B test, but one could still retain 65% of its sealing force when compressed while the other only retains 35%.

　　Often test temperatures are used for compression set tests that are substantially higher than those of field use. This can make the test data

压缩永久变形

橡胶部件
恒定负载/恒定变形
橡胶马达底座
O 形密封圈

汽车底盘

黏弹性的/存储能量
螺旋弹簧
消耗能量
黏性行为
重排/应力松弛

蠕变

载荷/变形
原始几何形状

一平方英寸/横截面

金属夹具

原始尺寸

永久变形

less meaningful in regard to the application. And in addition to that, in actual operation the O-rings are likely to undergo other changes due to contact with whatever fluid they are being used to seal, which again reduces the meaning of the laboratory test data in regard to real world use.

Compression set test <u>comparisons of</u> similar compounds will definitely have significance in deciding which is the best candidate for use in many applications, but when comparing dissimilar compounds there are many hazards in relying only on this kind of test. As with many other kinds of rubber tests, compression set testing is often used as much as a <u>quality control</u> tool as for evaluating fitness for use.

比较

质量控制

New Words

compression	[kəmˈpreʃn]	n.	压缩
set	[set]	adj.	处于某种状况；固定的
mount	[maʊnt]	n.	底座
chassis	[ˈʃæsi]	n.	（车辆的）底盘
seal	[siːl]	n.	密封
seal	[ˈsiːl]	v.	密封
dissipate	[ˈdɪsɪpeɪt]	v.	消散；挥霍
rearrangement	[ˌriːəˈreɪndʒmənt]	n.	重排，重新整理
indicator	[ˈɪndɪkeɪtə(r)]	n.	指示；指示者
retain	[rɪˈteɪn]	v.	保持，保留
display	[dɪˈspleɪ]	n.	展示，展现
fitness	[ˈfɪtnəs]	n.	合情理；合理性
comparison	[kəmˈpærɪsn]	n.	比较
hazard	[ˈhæzəd]	n.	冒险的事；危害物

Notes

[1] In both cases it would be desirable for the rubber to continue to fulfill its function in exactly the same way over time; that way the position of the engine would not change in relation to the car chassis, and the O-ring would exert the same sealing force after months or years of service as it did when first installed. 在这两种情况下，都希望橡胶持续的以完全相同的方式实现其功能；这样发动机的位置相对于汽车底盘不会发生变化，并且O形圈在经过几个月或几年的使用后依然具有与第一次安装时相同的密封性能。

[2] But elastomers are viscoelastic, which means they will store energy reversibly (elasticity) as a coiled spring does, but also that they will dissipate energy through viscous behavior, like a shock absorber. 但弹性体是黏弹性的，这意味着它们能像螺旋弹簧一样可逆地存储能量（弹性），而且它们将通过黏性特性消耗能量，如减振器一般。

[3] In practice, this guarantees that overtime the motor mount will deflect a little more (drift or creep), allowing the engine to move closer to the ground, and the O-ring will exert less sealing force than originally. In either case, if the rubber part is freed from its load/deformation,

it will not entirely regain its original geometry. 在实践中，这使得马达底座在长时间作用下会偏移一点点（漂移或蠕变），从而使发动机更靠近地面；O 形圈会施加比原来更小的密封力。在任何一种情况下，如果橡胶部件不受其载荷/变形的影响，就不会完全恢复其原始几何形状。

[4] The difference between the recovered dimensions of the rubber and its original dimensions is referred to as set, and the amount of set is an indicator of how much stress relaxation the rubber underwent while subjected to the load or the deformation. 橡胶试样恢复后的尺寸与其原始尺寸之间的差异被称为变形，变形的值能表明橡胶在经受负载或变形后应力松弛的程度。

[5] In the case of the constant imposed deformation, if the rubber did not recover at all but retained the dimensions imposed during the test, that would be a set of 100%, which would of course correlate with complete stress relaxation. 在持续恒定变形的情况下，如果橡胶不能完全恢复，而是保持在测试过程中的尺寸，那么它的（该橡胶试样的）压缩变形应当是 100%，这与完全的应力松弛吻合。

Exercises

1. Translate the following two paragraphs into Chinese.

Compression set is defined as the residual deformation of a material after removal of an applied compressive stress. Resistance to compression set is the ability of an elastomeric material to recover to its original thickness after having been compressed for an extended period.

Although there are differences of opinion about its validity, the compression set test is the easiest to run and therefore the most popular method of measuring the ability of a compound to return to its original shape and dimensions after being deformed for a long time. Low compression set data do not necessarily correlate with high resilience or low creep.

2. Put the following words and phrases into English.

| 压缩永久变形 | 恒定负载 | 恒定变形 | 黏弹性 | 蠕变 |
| 应力松弛 | 黏性行为 | 原始尺寸 | 质量控制 | O 形密封圈 |

[Reading Material]

Compression Set Testing

Compression set tests may be run either by applying a known constant force to the test specimen (Method A of ASTM D395) or by compressing the specimen to a known constant deflection (Method B of ASTM D395). The specimen is held in the compressed position for a specified period of time at a specified temperature, after which the compressive force is removed and the specimen is allowed to recover at room temperature for 30 minutes. In Method A, compression set is the difference between the original and final thickness of specimen, expressed as a percentage of the original thickness. In Method B, compression set is calculated by expressing the difference between the original and final thickness of the specimen as a percentage of the deflection to which the specimen was subjected. The standard specimen for a compression set test is a round pellet. O-rings of specified dimensions can also be tested.

Compression set tests are usually run at elevated temperatures, to simulate conditions or aging effects. Common test conditions are 70 hours at 70℃ or 100℃, although heat-resistant materials such as fluoro-elastomers may be tested for longer periods of time at temperatures up to

200℃ or more. If the end product is expected to perform at low temperatures, e.g., below 0℃, compression set is measured at the expected service temperature.

For optimum performance in service, compression set values should be as low as possible. Low set values mean that the material has recovered nearly to its original height, and there is very little residual deformation. This is particularly important in applications where a rubber part is expected to provide a seal under a compressive force, and the sealing force is removed and reapplied repeatedly.

Words and Expressions

specimen	['spesɪmən]	n.	样品;试样
deflection	[dɪ'flekʃn]	n.	变形;偏斜
specified temperature			指定温度
compressive force			压缩力
heat-resistant material			耐热材料
residual deformation			残余变形

Lesson 41　Abrasion Resistance

[1]Abrasion resistance is defined as the resistance of a rubber composition to wearing away by contact with a moving abrasive surface. It is usually reported as an abrasion resistance index, which is a ratio of the abrasion resistance of the test compound compared to that of a reference standard measured under the same test conditions.

Abrasion resistance is measured under definite conditions of load, speed and type of abrasive surface. The standard laboratory tests generally cannot be used to predict service life, because factors affecting abrasion are complex and vary greatly from application to application. Nevertheless, abrasion resistance tests are useful in making quality control checks on rubber products intended for rough service.

The most frequently used abrasion tester is the National Bureau of Standards (NBS) abrader. Test samples are pressed under a specified load against a rotating drum covered with abrasive paper. [2]The number of revolutions of the drum required to wear away a specified thickness of the test specimen is recorded and compared to the number of revolutions required to wear away the same thickness of standard reference material (ASTM D1630).

The Pico Abrader is used for measuring abrasion resistance of soft vulcanized rubber compounds and other elastomeric materials. A pair of tungsten carbide knives of specified geometry and controlled sharpness are rubbed over the surface of a pellet-shaped sample in a rotary fashion under controlled conditions of load, speed and time. The volume loss of the test material is measured and compared to a reference compound (ASTM D2228).

The Taber Abrader measures wear by weight and/or thickness loss. Specified abrasive wheels are held under load of 500 or 1000 grams against a test specimen mounted on a rotating turntable. The equipment is described in ASTM D1044, but the complete method is not followed for elastomer evaluations.

耐磨性

耐磨指数

使用寿命

质量控制检查

美国国家标准局磨耗仪
旋转辊筒

软质硫化橡胶
碳化钨刀片/特定几何形状
体积损失

质量和/或厚度损失
指定的砂轮

New Words

abrasion	[əˈbreɪʒn]	n.	磨损
abrasive	[əˈbreɪsɪv]	adj.	摩擦的；粗糙的
reference	[ˈrefrəns]	n.	参考，参照
definite	[ˈdefɪnət]	adj.	明确的；一定的
abrader	[əˈbreɪdə]	n.	磨耗仪
drum	[drʌm]	n.	鼓状物

turntable	['tɜːnteɪbl]	*n.*	转盘

Notes

[1] Abrasion resistance is defined as the resistance of a rubber composition to wearing away by contact with a moving abrasive surface. 耐磨性被定义为橡胶通过与移动的研磨表面接触，抵抗磨损的能力。

[2] The number of revolutions of the drum required to wear away a specified thickeness of the test specimen is recorded and compared to the number of revolutions required to wear away the same thickness of standard reference material (ASTM D1630). 记录磨除磨耗试样指定厚度所需的辊筒转数，并与磨除相同厚度的标准参照材料所需的辊筒转数相比较（ASTM D1630).

Exercises

1. Translate the second paragraph of the text into Chinese.
2. Put the following words and phrases into English.

耐磨性	耐磨指数	使用寿命	美国国家标准局	磨耗仪
质量损失	体积损失	厚度损失		

[Reading Material]

Tear Testing

Tear strength or Tear resistance of rubber is defined as the maximum force required to tear a test specimen in a direction normal to (perpendicular to) the direction of the stress. Tear strength is expressed as force per unit of specimen thickness—pounds force per inch (lbf/in), kilograms force per centimeter (kgf/cm), or kiloNewtons per meter (kN/m).

There are several types of tear specimens, with no apparent correlation between types. Some are cut to provide a starting point for tearing, while others are not. The most frequently used sample is an un-nicked 90 degree angle specimen listed in ASTM D624 as Die C and sometimes referred to as Graves Tear. Other specimens used are a razor-nicked crescent specimen—Die A of ASTM D624, also referred to as Winkleman Tear and a trouser tear specimen, so called because it resembles a pair of men's trousers, of ASTM D470.

Tear test specimens are pulled on a tensile tester at a cross-head rate of 8.5 millimeters per second (20 inches per minute). The maximum force required to initiate or propagate tear is recorded as force per unit thickness. Tear strength is frequently used to indicate relative toughness of different compounds. However, it is difficult to correlate with end-use performance.

Words and Expressions

tear	[teə(r)]	*n.*	撕裂,裂口
tear strength			撕裂强度
tear resistance			抗撕裂性能
perpendicular	[ˌpɜːpənˈdɪkjələ(r)]	*adj.*	垂直的,成直角的

trouser tear specimen			裤型撕裂试样
resemble	[rɪˈzembl]	v.	与…相像，类似于
tensile tester			拉伸测试仪
toughness	[tʌfnəs]	n.	韧性

Lesson 42 Fluid Resistance

[1] Applications of rubber bring the material into contact with an infinite variety of fluids and conditions, from drinking water at ambient temperature to motor oil over 100℃ to exotic diesel based drilling muds at temperatures in excess of 150℃ and several atmospheres of pressure. Effects of fluids on rubber range from very little at all to reversible swelling with minimal effect on functionality to aggressive chemical attack that destroys the material within days or even hours. Therefore it can be critical to know how some combination of fluid and rubber interact with each other.

Fluid resistance is a general term describing the extent to which a rubber product retains its original physical characteristics and ability to function when it is exposed to oil, chemicals, water, organic fluids or any liquid which it is likely to encounter in actual service. Fluid resistance tests may not give a direct correlation with service performance, because service conditions cannot easily be defined or controlled. However, they yield comparative data on which to base judgements about expected performance, and they serve as useful tools for screening compounds for particular service conditions.

Fluid tests are simple in concept and require immersion of sample pieces in the chosen fluid at an appropriate time/temperature combination. [2] There are several standard fluids described in the method, intended to be typical of various kinds of commonly encountered oils and fuels, and these are very frequently used when compounds undergo qualification so as to provide data readily compared with historical data for other compounds. Suggested conditions of time/temperature also appear in the ASTM D471 method to further simplify comparisons.

Fluid resistance of rubber compounds is principally a function of the base polymer. Many kinds of effects of fluids on rubber can be observed, not all of which are necessarily of interest to the end user. [3] While in most cases the rubber will gain mass/volume from fluid exposure, sometimes mass/volume will decrease. Hardness usually decreases, but again, sometimes increases. Tensile properties can also vary in different ways.

At a minimum hardness and mass changes are reported, and often volume change, but other responses may also be examined, such as change in tensile strength, elongation, etc. Interestingly, mass and volume changes do not always correlate closely, and some technologists will prefer to use one over the other as their criterion for how much the material has been affected by the fluid. There are no rigid standards for what is excessive swell, but roughly speaking, fluids can have three levels of effect on rubber: minimal changes, that is volume change from -5%

to 20%; moderate change, volume change from 20% to 75% swell; and high swell, which can go as high as 200%.

Hardness changes follow the amount of swell, with major decrease in hardness accompanying any large volume increase. Changes in hardness greater than a 10 point increase or a 5 point decrease are often considered to be an objectionable level of fluid effect, as would be a loss of half the original compound elongation. However, there is a wide range of specification limits on what is permissible change of rubber properties after fluid exposure, so any generalization of what are good or poor results of fluid testing would be misleading.

There is usually a major difference between a rubber article truly <u>soaking in a fluid</u> and the same article being exposed to occasional splashes or mists of the fluid. There is also a very significant contrast between fluid permeating a lab specimen 0.080 inches thick and fluid slowly soaking into a large rubber part with a low ratio of surface to volume. Therefore, even if some compound were shown to be seriously affected after 70 hours of soaking in a kind of oil at 100℃, an actual part made of the material and subjected to irregular splashes of the same fluid at room temperature might display no significant change in properties or functionality for a very long time.

浸泡在流体介质中

New Words

fluid	['fluːɪd]	n.	液体，流体
swelling	['swelɪŋ]	n.	膨胀；增大
reversible	[rɪ'vɜːsəbl]	adj.	可逆的
aggressive	[ə'gresɪv]	adj.	侵略的，侵犯的
immersion	[ɪ'mɜːʃn]	n.	浸润，沉浸
qualification	[ˌkwɒlɪfɪ'keɪʃn]	n.	授权；合格证书
exposure	[ɪk'spəʊʒə(r)]	n.	暴露；接触
criterion	[kraɪ'tɪəriən]	n.	（判断等的）标准
objectionable	[əb'dʒekʃənəbl]	adj.	令人不快的，有害的
misleading	[ˌmɪs'liːdɪŋ]	adj.	误导性的
soaking	[ˌsəʊkɪŋ]	n.	浸湿，浸透
permeate	['pɜːmieɪt]	v.	弥漫；渗透

Notes

[1] Applications of rubber bring the material into contact with an infinite variety of fluids and conditions, from drinking water at ambient temperature to motor oil over 100℃ to exotic diesel based drilling muds at temperatures in excess of 150℃ and several atmospheres of pressure.

橡胶在应用中与各种条件下的流体介质接触，从环境温度下的饮用水到超过100℃的发动机油，再到温度超过150℃、压力为几个大气压的柴油钻井泥浆。

[2] There are several standard fluids described in the method, intended to be typical of various kinds of commonly encountered oils and fuels, and these are very frequently used when compounds undergo qualification so as to provide data readily compared with historical data for other compounds. 该方法中描述了几种标准溶液，它们是各种常见油和燃料的典型代表，当需要对胶料进行鉴定时经常使用这些标准溶液，以便与其他胶料的历史数据相比较。

[3] While in most cases the rubber will gain mass/volume from fluid exposure, sometimes mass/volume will decrease. 尽管在大多数情况下，橡胶与流体介质接触后其质量和体积会增加，但有时其质量/体积会降低。

Exercises

1. Translate the following paragraph into Chinese.

Determining fluid resistance involves measuring a specimen's weight, volume and physical properties before and after exposure to the selected fluids for a specified time at a specified temperature. The effect of the fluid on the specimen is judged on the basis of the percent of the original properties retained or lost. Properties most often checked are volume change, weight change, and changes in hardness, tensile strength and elongation at break. Changes in tear strength, compression set, and low temperature properties are also frequently evaluated.

2. Put the following words and phrases into English.

耐介质性	溶胀	可逆溶胀	标准溶液	硬度变化
质量变化	体积变化			

[Reading Material]

Oil Resistance

Oil resistance is a special case of fluid resistance. Broadly defined, oil resistance is the ability of a rubber product to perform its intended function while in contact with oil. This recognizes that the elastomer may be swollen or weakened to some extent by the oil.

The tire rubbers (natural rubber, SBR), when placed in oil, absorb the fluid slowly until either the oil is all gone or the rubber has disintegrated. They never reach equilibrium. The so-called oil-resistant elastomers absorb some oil, especially at elevated temperature, but only a limited amount. With some it may be negligible. Most end uses requiring oil-resistant elastomers can tolerate appreciable swelling or volume increase. Hence, volume increase is not a very significant way to measure the ability of a rubber article to perform its intended function in oil.

Words and Expressions

oil resistance	耐油性
intended function	预期功能

swollen	['swəulən]	adj.	膨胀的
equilibrium	[ˌiːkwɪ'lɪbriəm]	n.	平衡
oil-resistant elastomer			耐油橡胶
negligible	['neglɪdʒəbl]	adj.	可以忽略的

Lesson 43　Heat Resistance

The phrase "heat resistance" as used in the rubber industry really means resistance to irreversible changes in properties as a result of prolonged exposure to high temperature. A rubber vulcanizate heated to high temperature can undergo two different kinds of temporary or reversible change: expansion, and softening (plastic flow). [1] A rubber vulcanizate aged at high temperature can undergo irreversible chemical changes: further crosslinking, cleavage of polymer chains by oxidation, or loss of some ingredients by evaporation or migration. [2] Comparisons of heat resistance generally deal with the chemical stability of the vulcanized polymer in terms of crosslinking and oxidation. The current commercial elastomers differ greatly in inherent heat resistance.

The normal means of measuring heat resistance is to determine stress-strain properties before and after heat aging. Interpretation of results is done in various ways, depending on what the chemist or end user feels is most relevant. Heat resistance is expressed in several ways:

➢ Percent change in elongation caused by heat aging.
➢ Percent change in tensile strength caused by heat aging.
➢ Points change in hardness caused by heat aging.
➢ Percent of original elongation retained through heat aging.
➢ Time of aging in hours until breaking elongation drops to 100%.

In an elastomer which hardens during heat aging, loss of rubber-like flexibility is equated with reduction in elongation.

New Words

prolonged	[prəˈlɒŋd]	adj.	持续很久的
temporary	[ˈtemprəri]	adj.	临时的;短暂的
expansion	[ɪkˈspænʃn]	n.	膨胀
softening	[ˈsɒfnɪŋ]	n.	软化,变软
cleavage	[ˈkliːvɪdʒ]	n.	分裂
evaporation	[ɪˌvæpəˈreɪʃn]	n.	蒸发
migration	[maɪˈgreɪʃn]	n.	迁移
inherent	[ɪnˈhɪərənt]	adj.	固有的;天生
interpretation	[ɪnˌtɜːprɪˈteɪʃn]	n.	解释,说明
retain	[rɪˈteɪn]	v.	保存,保持

Notes

[1] A rubber vulcanizate aged at high temperature can undergo irreversible chemical changes: further crosslinking, cleavage of polymer chains by oxidation, or loss of some ingredients by evaporation or migration. 在高温下老化的橡胶硫化胶会发生不可逆化学变化：进一步交

联，通过氧化使橡胶分子链裂解，或通过蒸发或迁移而损失一些配合剂。

[2] Comparisons of heat resistance generally deal with the chemical stability of the vulcanized polymer in terms of crosslinking and oxidation. 耐热性的比较通常涉及硫化聚合物在交联和氧化方面的化学稳定性。

Exercises

1. Translate the following two paragraphs into Chinese.

The technique used is to measure original tensile properties on specimens, then expose a matching set of specimens to heat over time, measure their tensile properties afterwards, and calculate the changes in properties resulting from the heat aging. Almost all the time the elongation decreases and tensile moduli increase, and usually tensile strength decreases. There are no scientifically justified criteria for what level of property change is acceptable or unacceptable in general, although obviously the less the properties change the better.

The particular combination of time and temperature used depends largely on the type of polymer being tested. A natural rubber formulation might be exposed to 70℃ for 22 hours, while for a silicone compound, 70 hours at 200℃ would be more appropriate. Generally speaking, the heat resistance of any rubber compound is primarily a function of the base polymer used.

2. Put the following words and phrases into English.

| 耐热性 | 化学稳定性 | 不可逆变化 | 化学变化 | 塑性流动 |
| 应力-应变特性 | 伸长率变化百分比 | 拉伸强度 | 膨胀 | 迁移 |

[Reading Material]

Low-Temperature Properties

As elastomer compositions are cooled below room temperature, they stiffen and become more difficult to bend, twist, or stretch. This stiffening is gradual until the stiffening temperature (also called the second order transition temperature) is reached. Then, further decrease in temperature causes a very sharp increase in stiffness. At very low temperatures, the elastomer composition becomes brittle and will crack or break if subjected to sudden impact or bending; the temperature at which this occurs is known as the brittle point or brittleness temperature.

The brittle point has no relationship to stiffness as shown by stiffness vs. temperature curves. For example, stiffness measurements may indicate a high degree of flexibility at a certain temperature, but impact tests may show the composition to be brittle at the same or higher temperature. This lack of relationship is not surprising, however, because stiffness measurements involve loading at low speed and low deflection, while brittleness tests involve loading at high speed and high deflection.

Words and Expressions

| stiffen | ['stɪfn] | v | 变硬 |
| second order transition temperature | | | 二阶转变温度 |

stiffening temperature			硬化温度
sharp increase in stiffness			硬度急剧增加
brittle	[ˈbrɪtl]	adj.	易碎的
brittleness temperature			脆性温度

Lesson 44　Ozone Resistance 耐臭氧性

　　The aggressive attack by atmospheric ozone on many widely used types of rubber can be a <u>limiting factor</u> 限制因素 in rubber product life. [1] In worst cases the rubber surface undergoes cracking that proceeds all the way to deep fissures and even rupture of thin sections. The demonstration of at least some level of ozone resistance is therefore a common requirement for compounds. [2] Exceptions to this include an O-ring which is to spend its working life sealed away from the atmosphere and has little need for ozone resistance, or the <u>undersea elastomeric bearings</u> 海底弹性体轴承 supporting the legs of an oil platform.

　　Ozone resistance tests all involve the exposure of the rubber to some level of ozone in air, typically a <u>partial pressure</u> 分压 of 50 or 100MPa, for some combination of time and temperature (usually 40℃ or 50℃), during or after which the sample pieces are examined for the presence of <u>cracking</u> 龟裂. [3] Putting rubber into a state of strain makes it dramatically more vulnerable to ozone's attack, and the higher the strain, the more the attack is accelerated. This is why different methods will call out different <u>specimen geometries</u> 试样几何形状 and different levels of static or <u>dynamic strain</u> 动态应变, achieved by stretching or bending, etc. The kinds of changes in sample strain achieved by these different methods will produce substantial contrasts in the rate of ozone attack, which is usually described by time to initial cracking.

　　By means of these tests it is certainly possible to validly compare compounds for their <u>relative ozone resistance</u> 相对耐臭氧性. However, <u>extrapolation</u> 推断 of survival time in the ozone test chamber to long term life predictions is <u>approximate</u> 近似的 at best.

New Words

fissure	['fɪʃə(r)]	n.	狭长裂缝
vulnerable	['vʌlnərəbl]	adj.	易受攻击的
extrapolation	[ɪkˌstræpə'leɪʃn]	n.	推断
prediction	[prɪ'dɪkʃn]	n.	预言,预测
approximate	[ə'prɒksɪmət]	adj.	大概的,近似的

Notes

　　[1] In worst cases the rubber surface undergoes cracking that proceeds all the way to deep fissures and even rupture of thin sections. 在最糟糕的情况下，橡胶表面会发生开裂直至发展成为深裂缝甚至破裂。

　　[2] Exceptions to this include an O-ring which is to spend its working life sealed away from the atmosphere and has little need for ozone resistance, or the undersea elastomeric bearings sup-

porting the legs of an oil platform. 例外的情况包括一个 O 形圈，它在使用过程中远离大气，并且几乎不需要具有耐臭氧性；或者支撑着石油平台支腿的海底弹性体轴承。

[3] Putting rubber into a state of strain makes it dramatically more vulnerable to ozone's attack, and the higher the strainis, the more the attack is accelerated. 将橡胶置于一定应变状态下，橡胶试样更容易受到臭氧的进攻，而且应变越高，进攻就越快。

Exercises

1. Translate the following two paragraphs into Chinese.

Weathering is a complex combination of erosion and UV light-induced oxidation. No elastomer is immune to this, especially if it is highly filled and light in color. Ozone attack, on the other hand, happens only to vulcanizates of unsaturated polymers when ozone is present in the atmosphere. Saturated elastomers, like EPDM, are virtually immune to ozone attack.

Tests for weather and ozone resistance depend on visual observation of specimens (e.g., for the appearance or growth of checks or cracks), and are therefore somewhat inaccurate and qualitative. The differences are typically so gross, nevertheless, that they can be demonstrated dramatically. A given ozone containing atmosphere can cause visible attack on one vulcanizate in a few minutes and no attack on another in a week.

2. Put the following words and phrases into English.

耐臭氧性　　　　限制因素　　　　龟裂　　　　臭氧龟裂　　　　动态应变

[Reading Material]

Deterioration by Ozone

Ozone is a highly reactive gas which can cause rubber products to crack and fail prematurely unless they are protected by antiozonants or made of an ozone-resistant elastomer. To test for resistance to ozone attack, samples are stretched to 20% or 40% strain on a test rack. The specimens are then placed in the ozone chamber, a special oven equipped with an ozone generator. Ozone concentration in the chamber can be controlled at any desired level—usual test concentrations are 0.5, 1, 3, or 100 ppm (parts ozone per million parts of air, by volume). Test temperature is usually 40℃. The test specimens are inspected at various time intervals until initial cracking occurs. Testing may continue until the specimen actually breaks, or may be stopped at some pre-determined degree of attack.

Dynamic testing provides a more rigorous test for ozone attack. Here, a special fabric-backed specimen is continuously flexed over a roller within the ozone chamber. Any protective chemical films which might build up on the surface of the specimen in static testing are quickly broken by the continuous flexing in dynamic testing. Interpretation of results is the same as for static testing. ASTM D-1149 covers static testing and ASTM D-3395 covers dynamic testing in a controlled ozone atmosphere.

Summary

The testing of rubbery materials and components made of rubber presents a large and complex field, which continues to grow with the increasing demands of modern industry and quality control practice. Not all tests hold as much meaning as might be implied, and careful consideration must be given as to which standard tests should be run for a particular situation and the meaning of the resulting data. Sometimes specialize tests may have to be developed in order to properly address concerns for suitability of a compound or component for use in a demanding application.

This has been an introductory treatment of a subject which has a great deal of depth, and there are many other sources of information on rubber testing for those who wish to pursue greater understanding.

Words and Expressions

rack	[ræk]	n.	支架
test rack			测试架
ozone generator			臭氧发生器
interval	[ˈɪntəvl]	n.	间隔
time intervals			时间间隔
rigorous	[ˈrɪɡərəs]	adj.	缜密的,严格的
quality control			质量控制
particular situation			特定情况
suitability	[ˌsjuːtəˈbɪlətɪ]	n.	合适,合适性
introductory	[ˌɪntrəˈdʌktəri]	adj.	引导的,介绍性的

References

[1] Hofmann W. Rubber Technology Handbook. Munich, Germany: Hanser Publishers, 1989: 2.
[2] Rubber & Plastics News, 1984, (14) 2: 21.
[3] Blow C M, in Rubber Technology and Manufacture. London: Newnes-Butterworths, 1977: 29.
[4] Heinisch K E. Dictionary of Rubber. New York: John Wiley & Sons, Inc., 1974: 451.
[5] Medalia A I, Juengel R R, Collins J M. Developments in Rubber Technology-1, Ed., A. Whelan and K. S. Lee. London: Applied Science Publishers Ltd., 1979: 163.
[6] Byers J T. Rubber Technology, Ed., Morton M. New York: Van Nostrand Reinhold, 1987: 56-61.
[7] Colombian Chemicals. The Nature of Carbon Black, November 1990: 16.
[8] ASTM D 1765-98 Standard Classification System for Carbon Blacks Used in Rubber Products.
[9] McCaffrey E C, Church E C and Jones F E. Profile of Carbon Blacks in Styrene-Butadiene Rubber, Technical Report RG-129, Cabot Corporation.
[10] Fetterman M Q. Rubber World, 1986, (194) 1: 38.
[11] Shelton J R. Rubber Chem. Technol., 1972, 45, 359.
[12] Criegee R, Blust G, Zinke H. Chem. Ber., 1954, 87, 766.
[13] Andries J C, Ross D B, Diem H E. Rubber Chem. Technol., 1975, 48, 41.
[14] Morris G. Developments in Rubber Technology-1, Ed., Whelan A and Lee K S. London: Applied Science Publishers Ltd., 1979: 219.
[15] "Basic Rubber Compounding and Processing of Rubber," Rubber Division, ACS, Long H, Ed., 1985, ch. 2: 32.
[16] Freakley P K. Rubber Processing and Production Organization. [New York: Plenum Press, 1985 (ISBN 0-306-41745-6).
[17] Grossman R F, Eds. The Mixing of Rubber. London: Chapman and Hall, 1997 (ISBN 0-412-80490-5).
[18] Turk K J. in Injection Molding of Elastomer. Penn W S, Ed.. London: Maclaren and Sons, 1968, Ch. 3.
[19] Willshaw H. Calenders for Rubber Processing. London, England: The Institution for Rubber Processing, Lakeman & Co., 1956: 3-5.
[20] Ohm R F. The Vanderbilt Rubber Handbook. Norwalk, CT: R. T. Vanderbilt Co., 1990.
[21] Morton M. Rubber Technology. New York: Van Nostrand-Reinhold, 1987.
[22] Morton M. Introduction to Rubber Technology. New York: Reinhold, 1959.
[23] Bateman L. The Chemistry and Physics of Rubber-Like Substances. [New York: Wiley, 1963.
[24] Dick J S. Compounding Materials for the Polymer Industries. Park Ridge, NJ: Noyes, 1987.
[25] Stephens H L. Textbook for Intermediate Correspondence Course. Washington DC: American Chemical Society, Rubber Division, 1985.
[26] Katz H S, Milewski J V. Handbook of Fillers for Plastics. New York: Van Nostrand-Reinhold, 1987.
[27] Hagemeyer R W. Pigments for Paper. Atlanta: TAPPI Press, 1984.
[28] Eirich F R. Science and Technology of Rubber. New York: Academic Press, 1978.
[29] Mark H F, et al. Encyclopedia of Polymer Science and Technology. New York: Interscience, 1971.
[30] Norman R H, Johnson P S. Rubber Chem. Technol, 1991, 54, 493.
[31] Mooney M. Ind. Eng Chem., Anal. Ed, 1934, 6, 147.
[32] ASTM D1646-98, "Standard Test Method for Rubber-Viscosity, Stress Relazation, and Pre-Vulcanization Characteristics (Mooney Viscometer)," ASTM, 1916 Race Street, Philadelphia, PA19103-1187.